50 Leveled Math Problems

150 Problems Total

Author
Anne M. Collins, Ph.D.

Contributing Author and Consultant

Linda Dacey, Ed.D.
Professor, Lesley University

Consultant

Kaitlyn Aspell
Achievement Center for Mathematics
Lesley University

Publishing Credits

Dona Herweck Rice, *Editor-in-Chief*; Robin Erickson, *Production Director*; Lee Aucoin, *Creative Director*; Timothy J. Bradley, *Illustration Manager*; Sara Johnson, M.S.Ed., *Senior Editor*; Aubrie Nielsen, M.S.Ed., *Associate Education Editor*; Jennifer Kim, M.A.Ed., *Associate Education Editor*; Leah Quillian, *Assistant Editor*; Grace Alba, *Interior Layout Designer*; Ana Clark, *Illustrator*; Corinne Burton, M.A.Ed., *Publisher*

Standards
© 2003 National Council of Teachers of Mathematics (NCTM)
© 2004 Mid-continent Research for Education and Learning (McREL)
© 2007 Teachers of English to Speakers of Other Languages, Inc. (TESOL)
© 2010 National Governors Association Center for Best Practices and Council of Chief State School Officers (CCSS)

Shell Education
5301 Oceanus Drive
Huntington Beach, CA 92649-1030
http://www.shelleducation.com
ISBN 978-1-4258-0777-1
© 2012 Shell Educational Publishing, Inc.

The classroom teacher may reproduce copies of materials in this book for classroom use only. The reproduction of any part for an entire school or school system is strictly prohibited. No part of this publication may be transmitted, stored, or recorded in any form without written permission from the publisher.

Table of Contents

Introduction
Problem Solving in Mathematics Instruction 5
Understanding the Problem-Solving Process 7
Problem-Solving Strategies 12
Ask, Don't Tell 14
Differentiating with Leveled Problems 15
Management and Assessment 18
How to Use This Book 22
Correlations to Standards 26

Leveled Problem-Solving Lessons

Operations and Algebraic Thinking
In What Order? 32
Order Counts 34
Number Patterns 36
Geometric Patterns 38
How Else Might I Look? 40
Where Am I? 42
How Do I Change? 44
What's Our Relation? 46

Number and Operations in Base Ten
Name My Number 48
Rectangular Products 50
Whatever Remains 52
Grouping or Sharing? 54
Dealing with Decimals 56
Expanded Form 58
Travel Expenses 60
Computing with Decimals 62
Dizzying Decimals 64
Estimating Decimals 66
About How Much? 68

Number and Operations—Fractions
Where Do I Fit? 70
Ribbons and Bows 72
Fractional Sums 74

Table of Contents (cont.)

What's the Difference? 76
It's Close to What? 78
More or Less 80
Fractional Areas 82
The Product Is Smaller 84
Fair Sharing, Equal Groups 86
Map Reading 88

Measurement and Data

Fill It Up 90
How Spacious Is It? 92
Cubic Views 94
Volume in Practice 96
What's My Unit? 98
Metrically Speaking 100
How Much Is There? 102
All in a Line 104
What Is the Favorite? 106
Stem-and-Leaf Plots 108
What Does It Mean? 110
The Plot Thickens 112
One or the Other 114

Geometry

Congruency 116
Classifying Figures 118
Plots A Lot 120
Flips, Slides, and Turns 122
What Is the Angle? 124
Sort It Out 126
Geometric Nets 128
Graph It 130

Appendices

Appendix A: Student Response Form 132
Appendix B: Observation Form 133
Appendix C: Record-Keeping Chart 134
Appendix D: Answer Key 135
Appendix E: References Cited 142
Appendix F: Contents of the Teacher Resource CD 143

Problem Solving in Mathematics Instruction

If you were a student in elementary school before the early 1980s, your education most likely paid little or no attention to mathematical problem solving. In fact, your exposure may have been limited to solving word problems at the end of a chapter that focused on one of the four operations. After a chapter on addition, for example, you solved problems that required you to add two numbers to find the answer. You knew this was the case, so you just picked out the two numbers from the problem and added them. Sometimes, but rarely, you were assigned problems that required you to choose whether to add, subtract, multiply, or divide. Many of your teachers dreaded lessons that contained such problems as they did not know how to help the many students who struggled.

If you went to elementary school in the later 1980s or in the 1990s, it may have been different. You may have learned about a four-step model of problem solving and perhaps you were introduced to different strategies for finding solutions. There may have been a separate chapter in your textbook that focused on problem solving and two-page lessons that focused on particular problem-solving strategies, such as guess and check. Attention was given to problems that required more than one computational step for their solution, and all the information necessary to solve the problems was not necessarily contained in the problem statements.

One would think that the ability of students to solve problems would improve greatly with these changes, but that has not been the case. Research provides little evidence that teaching problem solving in this isolated manner leads to success (Cai 2010). In fact, some would argue that valuable instructional time was lost exploring problems that did not match the mathematical goals of the curriculum. An example would be learning how to use logic tables to solve a problem that involved finding out who drank which drink and wore which color shirt. Being able to use a diagram to organize information, to reason deductively, and to eliminate possibilities are all important problem-solving skills, but they should be applied to problems that are mathematically significant and interesting to students.

Today, leaders in mathematics education recommend teaching mathematics in a manner that integrates attention to concepts, skills, and mathematical reasoning. Referred to as *teaching through problem solving*, this approach suggests that problematic tasks serve as vehicles through which students acquire new mathematical concepts and skills (D'Ambrosio 2003). Students apply previous learning and gain new insights into mathematics as they wrestle with challenging tasks. This approach is quite different from introducing problems only after content has been learned.

Most recently, the *Common Core State Standards* listed the need to persevere in problem solving as the first of its Standards for Mathematical Practice (National Governors Association Center for Best Practices and Council of Chief State School Officers 2010):

> **Make sense of problems and persevere in solving them.**
>
> *Mathematically proficient students start by explaining to themselves the meaning of a problem and looking for entry points to its solution. They analyze givens, constraints, relationships, and goals. They make conjectures about the form and meaning of the solution and plan a solution pathway rather than simply jumping into a solution attempt. They consider analogous problems, and try*

Problem Solving in Mathematics Instruction (cont.)

special cases and simpler forms of the original problem in order to gain insight into its solution. They monitor and evaluate their progress and change course if necessary. Older students might, depending on the context of the problem, transform algebraic expressions or change the viewing window on their graphing calculator to get the information they need. Mathematically proficient students can explain correspondences between equations, verbal descriptions, tables, and graphs or draw diagrams of important features and relationships, graph data, and search for regularity or trends. Younger students might rely on using concrete objects or pictures to help conceptualize and solve a problem. Mathematically proficient students check their answers to problems using a different method, and they continually ask themselves, "Does this make sense?" They can understand the approaches of others to solving complex problems and identify correspondences between different approaches.

This sustained commitment to problem solving makes sense; it is the application of mathematical skills to real-life problems that makes learning mathematics so important. Unfortunately, we have not yet mastered the art of developing successful problem solvers. Students' performance in the United States on the 2009 Program for International Student Assessment (PISA), a test that evaluates 15-year-old students' mathematical literacy and ability to apply mathematics to real-life situations, suggests that we need to continue to improve our teaching of mathematical problem solving. According to data released late in 2010, students in the U.S. are below average (National Center for Educational Statistics 2010). Clearly we need to address this lack of success.

Students do not have enough opportunities to solve challenging problems. Further, problems available to teachers are not designed to meet the individual needs of students. Additionally, teachers have few opportunities to learn how best to create, identify, and orchestrate problem-solving tasks. *50 Leveled Math Problems* is a unique series that is designed to address these concerns.

Understanding the Problem-Solving Process

George Polya is known as the father of problem solving. In his book *How to Solve It: A New Aspect of Mathematical Method* (1945), Polya provides a four-step model of problem solving that has been adopted in many classrooms: understanding the problem, making a plan, carrying out the plan, and looking back. In some elementary classrooms this model has been shortened to: understand, plan, do, check. Unfortunately, this over-simplification ignores much of the richness of Polya's thinking.

Polya's conceptual model of the problem-solving process has been adapted for use at this level. Teachers are encouraged to view the four steps as interrelated, rather than only sequential, and to recognize that problem-solving strategies are useful at each stage of the problem-solving process. The model presented here gives greater emphasis to the importance of communicating and justifying one's thinking as well as to posing problems. Ways in which understanding is deepened throughout the problem-solving process is considered in each of the following steps.

Step 1: Understand the Problem

Students engage in the problem-solving process when they attempt to *understand the problem*, but the understanding is not something that just happens in the beginning. At grade 5, students may be asked to restate the problem in their own words and then turn to a neighbor to summarize what they know and what they need to find out. Students should also discuss the problem-solving strategies they might use. At this grade level, it is important for students to be able to identify which strategies would be appropriate to use for a given problem.

What is most important is that teachers do not teach students to rely on key words or show students "tricks" or "short-cuts" that are not built on conceptual understanding. Interpreting the language of mathematics is complex, and terms that are used in mathematics often have different everyday meanings. Note how a reliance on key words would lead to failure when solving the problem below. A student taught that *of* means *multiply* may multiply the fractions to find that one-half of the students (17 students) is the correct answer to the problem.

> *There are 18 boys and 16 girls in a classroom. Some are seated and some are standing. $\frac{2}{3}$ of all the boys and $\frac{3}{4}$ of all the girls are seated. How many students are seated?*

Step 2: Apply Strategies

Once students have a sense of the problem they can begin to actively explore it. They may do so by applying one or more of the following strategies. Note that we have combined related actions within some of the strategies.

- Act it out or use manipulatives.
- Count, compute, or write an equation.
- Find information in a picture, list, table, graph, or diagram.
- Generalize a pattern.
- Guess and check or make an estimate.
- Organize information in a picture, list, table, graph, or diagram.
- Simplify the problem.
- Use logical reasoning.
- Work backward.

Understanding the Problem-Solving Process (cont.)

Step 2: Apply Strategies (cont.)

As students apply these strategies, they also deepen their understanding of the *mathematics* of the problem. As such, understanding develops throughout the problem-solving phases. Consider the following problem requiring students to find the greatest common factor.

> Mrs. Freedman has 9 juice boxes and 15 granola bars. For his sleepover, she told her son that he could only bring home the number of friends who could evenly share the juice boxes and granola bars. The friends can have more than one juice box and one granola bar. How many friends might he invite?

> Riley and Isabelle are partners. They have read the problem and understand that they have to determine the greatest common factor. Riley suggests they make an organized list but Isabelle says she wants to use a Venn diagram. The girls decide they will do both and see if they get the same answer. Riley writes $9 = 3 \times 3$ and $15 = 5 \times 3$. She then crosses out a three in each set of factors. When her teacher asks her why she crossed out those threes, Riley answers, "Because they are the same." The teacher then asks her if she can cross out the other three. Riley responds, "Sure, you can cross out as many as you want." Isabelle makes her Venn diagram and shows the intersection of the factors of 9 and 15 only has a 3 in it. She explains to Riley that she cannot cross out any factor she wants, but instead needs to identify only factors that are shared.

The girls are at different stages of understanding, but the pairing of the two students is effective in helping Riley think about why she performs certain procedures and is exposing her to an additional strategy.

It is important that we offer students problems that can be solved in more than one way. If one strategy does not lead to success, students can try a different one. This option gives students the opportunity to learn that getting "stuck" might just mean that a new approach should be considered. When students get themselves "unstuck" they are more likely to view themselves as successful problem solvers. Such problems also lead to richer mathematical conversations as there are different ideas and perspectives to discuss. Consider the following problem:

> There are twenty-five students in Mr. Stewart's grade 5 class. If each student gives each classmate a high-five, how many high-fives will be exchanged?

Understanding the Problem-Solving Process *(cont.)*

Notice that in Student Sample 1, the student started by listing all the students in the class but crossed it out. This student then *broke the problem into a simpler problem* and *made an organized list*. The combination of these two strategies appears to help this student generalize the problem to the twenty-five students in the class. The student generalized the fact that he is adding one less than the number of students and has shown recognition of the need to add consecutive integers. This student then made a list of the integers from 1–24 and added them together for a sum of 300.

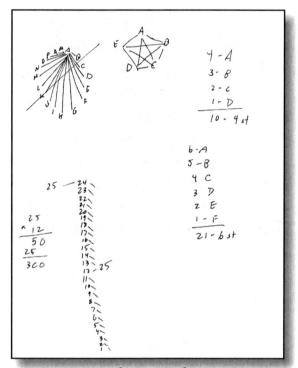

Student Sample 1

A second student made a table, identified a pattern, and also arrived at a sum of 300.

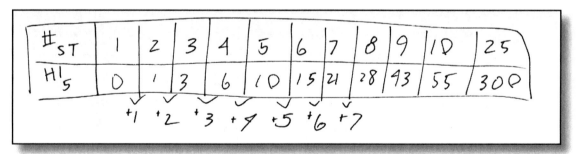

Student Sample 2

It is also important that students make records as they work so that they can recall their thinking.

Introduction

Understanding the Problem-Solving Process *(cont.)*

Step 3: Communicate and Justify Your Thinking

It is essential that teachers ask students to communicate and justify their thinking. It is also important that students make records as they work so that they can recall their thinking. When teachers make it clear that they expect such behavior from students, they are establishing an important habit of mind (Goldenberg, Shteingold, and Feurzeig 2003) and developing their understanding of the nature of mathematics. When students explain their thinking orally while investigating a problem with a partner or small group, they may deepen their understanding of the problem or recognize an error and fix it. When students debrief after finding solutions, they learn to communicate their thinking clearly and in ways that give others access to new mathematical ideas. In one class a fifth-grader listens to a peer's explanation and proclaims, *"Oh, I see. The common denominators will always simplify to one when we divide fractions using common denominators."* Such discourse is essential to the mathematical practice suggested in the *Common Core State Standards* that students "construct viable justifications and critique the reasoning of others" (National Governors Association Center for Best Practices and Council of Chief State School Officers 2010).

Our task is to foster learning environments where students engage in this kind of "accountable talk." Michaels, O'Connor, and Resnick (2008) identify three aspects of this type of dialogue. The first is that students are accountable to their learning communities; they listen to each other carefully and build on the ideas of others. Second, accountable talk is based on logical thinking and leads to logical conclusions. Finally, these types of discussions are based on facts or other information that is available to everyone.

When we emphasize the importance of discussions and explanations, we are teaching our students that it is the soundness of their mathematical reasoning that determines what is correct, not merely an answer key or a teacher's approval. Students learn, therefore, that mathematics makes sense and that they are mathematical sense-makers.

Understanding the Problem-Solving Process (cont.)

Step 4: Take It Further

Debrief

It is this final step in the problem-solving process to which teachers and students are most likely to give the least attention. When time is given to this step, it is often limited to *check your work*. In contrast, this step offers rich opportunities for further learning. Students might be asked to solve the problem using a different strategy, or to find additional solutions. They might be asked to make a mathematical generalization based on their investigation. Students might connect this problem to another problem they have solved already, or they now may be able to solve a new, higher-level problem.

Posing Problems

Students can also take problem solving further by posing problems. In fact, problem posing is intricately linked with problem solving (Brown and Walter 2005). When posing their own problems, students can view a problem as something they can create, rather than as a task that is given to them. This book supports problem posing through a variety of formats. For example, students may be asked to supply missing data in a problem so that it makes sense. They may be given a problem with the question omitted and asked to compose one. Or, they may be given both problem data and the answer and asked to identify the missing question. Teachers may also choose to ask students to create their own problems that are similar to those they have previously solved. Emphasis on problem posing can transform the teaching of problem solving and build lifelong curiosity in students.

Consider the following reflections of Renee, a fifth-grade teacher, after she explored *Grouping or Sharing?* (see page 54) with her students.

> My students continue to surprise me with their deep thinking and development of problem-solving skills. Most try really hard to phrase a conjecture as they work to predict, then prove, a mathematical idea. This only works because of the rich problems that I am now able to assign.
>
> When we did the Grouping or Sharing? lesson and the students had to pose a division problem given three values, it was amazing listening to them construct a contextual situation in which the numbers fit and the answer made sense. So often during that lesson I heard mathematical language that I was not accustomed to hearing from my fifth-grade students. I plan on assigning problem posing on a regular basis now that I understand the power behind it.

Problem-Solving Strategies

Think of someone doing repair jobs around the house. Often that person carries a toolbox or wears a tool belt from task to task. Common tools such as hammers, screwdrivers, and wrenches are then readily available. The repair person chooses tools (usually more than one) appropriate for a particular task. Problem-solving strategies are the tools used to solve problems. Labeling the strategies allows students to refer to them in discussions and helps students recognize the wide variety of tools available for the solution of problems. The problems in this book provide opportunities for students to apply one or more of the following strategies:

Act It Out or Use Manipulatives

Students' understanding of a problem is greatly enhanced when they act it out. Students may choose to dramatize a situation themselves or use manipulatives to show the actions or changes that take place. If students suggest they do not understand a problem say something such as *Imagine this is a play. Show me what the actors are doing.*

Count, Compute, or Write an Equation

When students count, compute, or write an equation to solve a problem they are making a match between a context and a mathematical skill. Once the connection is made, students need only to carry out the procedure accurately. Sometimes writing an equation is a final step in the solution process. For example, students might work with manipulatives or draw pictures and then summarize their thinking by recording an equation.

Find Information in a Picture, List, Table, Graph, or Diagram

Too often problems contain all of the necessary information in the problem statement. Such information is never so readily available in real-world situations. It is important that students develop the ability to interpret a picture, list, table, graph, or diagram and identify the information relevant to the problem.

Generalize a Pattern

Some people consider mathematics the study of patterns, so it makes sense that the ability to identify, continue, and generalize patterns is an important problem-solving strategy. The ability to generalize a pattern requires students to recognize and express relationships. Once generalized, the student can use the pattern to predict other outcomes.

Guess and Check or Make an Estimate

Guessing and checking or making an estimate provide students with insights into problems. Making a guess can help students to better understand conditions of the problem; it can be a way to try something when a student is stuck. Some students may make random guesses, but over time, students learn to make more informed guesses. For example, if a guess leads to an answer that is too large, a student might next try a number that is less than the previous guess. Estimation can help students narrow their range of guesses or be used to check a guess.

Problem-Solving Strategies (cont.)

Organize Information in a Picture, List, Table, Graph, or Diagram

Organizing information can help students both understand and solve problems. For example, students might draw a number line or a map to note information given in the problem statement. When students organize data in a table or graph they might recognize relationships among the data. Students might also make an organized list to keep track of guesses they have made or to identify patterns. It is important that students gather data from a problem and organize it in a way that makes the most sense to them.

Simplify the Problem

Another way for students to better understand a problem, or perhaps get "unstuck," is to simplify it. Often the easiest way for students to do this is to make the numbers easier. For example, a student might replace four-digit numbers with single-digit numbers or replace fractions with whole numbers. With simpler numbers students often gain insights or recognize relationships that were not previously apparent, but that can now be applied to the original problem. Students might also work with 10 numbers, rather than 100, to identify patterns.

Use Logical Reasoning

Logical thinking and sense-making pervades mathematical problem solving. To solve problems students need to deduce relationships, draw conclusions, make inferences, and eliminate possibilities. Logical reasoning is also a component of many other strategies. For example, students use logical reasoning to revise initial guesses or to interpret diagrams. Asking questions such as *What else does this sentence tell you?* helps students to more closely analyze given data.

Work Backward

When the outcome of a situation is known, we often work backward to determine how to arrive at that goal. We might use this strategy to figure out what time to leave for the airport when we know the time our flight is scheduled to depart. A student might work backward to answer the question *What did Joey add to 79 to get a sum of 146?* or *If it took 2 hours and 23 minutes to drive a given route and the driver arrived at 10:17, at what time did the driver leave home?* Understanding relationships among the operations is critical to the successful use of this strategy.

Ask, Don't Tell

All teachers want their students to succeed, and it can be difficult to watch them struggle. Often when students struggle with a problem, a first instinct may be to step in and show them how to solve it. That intervention might feel good, but it is not helpful to the student. Students need to learn how to struggle through the problem-solving process if they are to enhance their understanding and reasoning skills. Perseverance in solving problems is listed under the mathematical practices in the *Common Core State Standards* and research indicates that students who struggle and persevere in solving problems are more likely to internalize the problem-solving process and build upon their successes. It is also important to recognize the fact that people think differently about how to approach and solve problems.

An effective substitution for telling or showing students how to solve problems is to offer support through questioning. George Bright and Jeane Joyner (2005) identify three different types of questions to ask, depending on where students are in the problem-solving process: (1) engaging questions, (2) refocusing questions, and (3) clarifying questions.

Engaging Questions

Engaging questions are designed to pique student interest in a problem. Students are more likely to want to solve problems that are interesting and relevant. One way to immediately grab a student's attention is by using his or her name in the problem. Once a personal connection is made, a student is more apt to persevere in solving the problem. Posing an engaging question is also a great way to redirect a student who is not involved in a group discussion. Suppose students are provided the missing numbers in a problem and one of the sentences reads *Janel is about _____ centimeters tall and rides her bicycle to school.* Engaging questions might include *What do you know about 100 centimeters? Are you taller or shorter than 100 centimeters?* The responses will provide further insight into how the student is thinking.

Refocusing Questions

Refocusing questions are asked to redirect students away from a nonproductive line of thinking and back to a more appropriate track. These questions often begin with the phrase *What can you tell me about…?* or *What does this number…?* Refocusing questions are also appropriate if you suspect students have misread or misunderstood the problem. Asking them to explain in their own words what the problem is stating and what question they are trying to answer is often helpful.

Clarifying Questions

Clarifying questions are posed when it is unclear why students have used a certain strategy, picture, table, graph, or computation. They are designed to help demonstrate what students are thinking, but can also be used to clear up misconceptions students might have. The teacher might say *I am not sure why you started with the number 10. Can you explain that to me?*

As teachers transform instruction from "teaching as telling" to "teaching as facilitating," students may require an adjustment period to become accustomed to the change in expectations. Over time, students will learn to take more responsibility and to expect the teacher to probe their thinking, rather than supply them with answers. After making this transition in her own teaching, one teacher shared a student's comment: "I know when I ask you a question that you are only going to ask me a question in response. But, sometimes the question helps me figure out the next step I need to take. I like that."

Differentiating with Leveled Problems

There are four main ways that teachers can differentiate: by content, by process, by product, and by learning environment. Differentiation by content involves varying the material that is presented to students. Differentiation by process occurs when a teacher delivers instruction to students in different ways. Differentiation by product asks students to present their work in different ways. Offering different learning environments, such as small group settings, is another method of differentiation. Students' learning styles, readiness levels, and interests determine which differentiation strategies are implemented. The leveled problems in this book vary aspects of mathematics problems so that students at various readiness levels can succeed. Mini-lessons include problems at three levels and ideas for differentiation. These are designated by the following symbols:

- ● lower-level challenge
- ■ on-level challenge
- ▲ above-level challenge
- ★ English language learner support

Ideally, students solve problems that are at just the right level of challenge—beyond what would be too easy, but not so difficult as to cause extreme frustration (Sylwester 2003; Tomlinson 2003; Vygotsky 1986). The goal is to avoid both a lack of challenge, which might leave students bored, as well as too much of a challenge, which might lead to significant anxiety.

Differentiating with Leveled Problems (cont.)

There are a variety of ways to level problems. In this book, problems are leveled based on the concepts and skills required to find the solution. Problems are leveled by adjusting one or more of the following factors:

Complexity of the Mathematical Language

The mathematical language used in problems can have a significant impact on their level of challenge. For example, negative statements are more difficult to interpret than positive ones. *It is not an even number* is more complex than *It is an odd number*. Phrases such as *at least* or *between* also add to the complexity of the information. Further, words such as *table*, *face*, and *plot* can be challenging since their mathematical meaning differs from their everyday uses.

Complexity of the Task

There are various ways to change the complexity of the task. One example would be the number of solutions that students are expected to identify. Finding one solution that satisfies problem conditions is less challenging than finding more than one solution, which is even less difficult than identifying *all* possible solutions. Similarly, increases and decreases in the number of conditions that must be met and the number of steps that must be completed change the complexity of a problem.

Changing the Numbers

Sometimes it is the size of the numbers that is changed to increase the level of mathematical skills required. A problem may be more complex when it involves fractions, decimals, or negative integers. Sometimes changes to the "friendliness" of the numbers are made to adapt the difficulty level. For example, if two problems involve fractions, one with common denominators is simpler than one with unlike denominators.

Amount of Support

Some problems provide more support for learners than others. Providing a graphic organizer or a table that is partially completed is one way to provide added support for students. Offering information with pictures rather than words can also vary the level of support. The inclusion of such supports often helps students to better understand problems and may offer insights on how to proceed. The exclusion of supports allows a learner to take more responsibility for finding a solution, and it may make the task appear more abstract or challenging.

Differentiating with Leveled Problems (cont.)

Differentiation Strategies for English Language Learners

Many English language learners may work at a high readiness level in many mathematical concepts, but may need support in accessing the language content. Specific suggestions for differentiating for English language learners can be found in the *Differentiate* section of the mini-lessons. Additionally, the strategies below may assist teachers in differentiating for English language learners.

- Allow students to draw pictures or provide oral responses as an alternative to written responses.
- Pose questions with question stems or frames. Example question stems/frames include: *What would happen if…?, Why do you think…?, How would you prove…?, How is _____ related to _____?, and Why is _____ important?*
- Use visuals to give context to questions. Add pictures or icons next to key words, or use realia to help students understand the scenario of the problem.
- Provide sentence stems or frames to help students articulate their thoughts. Sentence stems include: *This is important because…, This is similar because…, and This is different because….* Sentence frames include: *I agree with _____ because…, I disagree with _____ because…, and I think _____ because….*
- Partner English language learners with language-proficient students.

Introduction

Management and Assessment

Organization of the Mini-Lessons

The mini-lessons in this book are organized according to the domains identified in the *Common Core State Standards*, which have also been endorsed by the National Council of Teachers of Mathematics. At grade 5, these domains are *Operations and Algebraic Thinking, Number and Operations in Base Ten, Number and Operations—Fractions, Measurement and Data,* and *Geometry.* Though organized in this manner, the mini-lessons are independent of one another and may be taught in any order within a domain or among the domains. What is most important is that the lessons are implemented in the order that best fits a teacher's curriculum and practice.

Ways to Use the Mini-Lessons

There are a variety of ways to assign and use the mini-lessons, and they may be implemented in different lessons throughout the year. They can provide practice with new concepts or be used to maintain skills previously learned. The problems can be incorporated into a teacher's mathematics lessons once or twice each week, or they may be used to introduce extended or additional instructional periods. They can be used in the regular classroom with the whole class or in small groups. They can also be used to support Response to Intervention (RTI) and after-school programs.

It is important to remember that a student's ability to solve problems depends greatly on the specific content involved and may change over the course of the school year. Establish the expectation that problem assignment is flexible; sometimes students will be assigned to one level (circle, square, or triangle) and sometimes to another. On occasion, you may also wish to allow students to choose their own problems. Much can be learned from students' choices!

Students can also be assigned one, two, or all three of the problems to solve. Although leveled, some students who are capable of wrestling with complex problems need the opportunity to warm up first to build their confidence. Starting at a lower level serves these students well. Teachers may also find that students correctly assigned to a below- or on-level problem will be able to consider a problem at a higher level after solving one of the lower problems. Students can also revisit these problems, investigating those at the higher levels not previously explored.

Grouping Students to Solve Leveled Problems

A differentiated classroom often groups students in a variety of ways, based on the instructional goals of an activity or the tasks students must complete. At times, students may work in heterogeneous groups or pairs with students of varying readiness levels. Other activities may lend themselves to homogeneous groups or pairs of students who share similar readiness levels. Since the problems presented in this book provide below-level, on-level, and above-level challenges, you may wish to partner or group students with others who are working at the same readiness level.

Since students' readiness levels may vary for different mathematical concepts and change throughout a course of study, students may be assigned different levels of problems at different times. It is important that the grouping of students for solving leveled problems stay flexible. Struggling students who feel that they are constantly assigned to work with a certain partner or group may develop feelings of shame or stigma. Above-level students who are routinely assigned to the same group may become disinterested and cause behavior problems. Varying students' groups can help keep the activities engaging.

Management and Assessment (cont.)

Assessment for Learning

In recent years, increased attention has been given to summative assessment in schools. Significantly more instructional time is taken with weekly quizzes, chapter tests, and state-mandated assessments. These tests, although seen as tedious by many, provide information and reports about achievement to students, parents, administrators, and other interested stakeholders. However, these summative assessments often do not have a real impact on an individual student's learning. In fact, when teachers return quizzes and tests, many students look at the grade and if it is "good," they bring the assessment home. If it is not an acceptable grade, they often just throw away the assessment.

Research shows that to have an impact on student learning we should rely on assessments *for* learning, rather than on assessments *of* learning. That is, we should focus on assessment data we collect during the learning process, not after the instructional cycle is completed. These assessments for learning, or formative assessments, are shown to have the greatest positive impact on student achievement (National Mathematics Advisory Panel 2008). Assessment for learning is an ongoing process that includes a variety of strategies and protocols to inform the progression of student learning.

One might ask, "So, what is the big difference? Don't all assessments accomplish the same goal?" The answer to those key questions is *no*. A great difference is the fact that formative assessment is designed to make student thinking visible. This is a real transformation for many teachers because when the emphasis is on student thinking and reasoning, the focus shifts from whether the answer is correct or incorrect to how the students grapple with a problem. Making student thinking visible entails a change in the manner in which teachers interact with their students. For instance, instead of relying solely on students' written work, teachers gather information through observation, questioning, and listening to their students discuss strategies, justify their reasoning, and explain why they chose to make particular decisions or use a specific representation. Since observations happen in real time, teachers can react in the moment by making an appropriate instructional decision, which may mean asking a well-posed question or suggesting a different model to represent the problem at hand.

Students are often asked to explain what they were thinking as they completed a procedure. Their response is often a recitation of the steps that were used. Such an explanation does not shed any light on whether a student understands the procedure, why it works, or if it will always work. Nor does it provide teachers with any insight into whether a student has a superficial or a deep understanding of the mathematics involved. If, however, students are encouraged to explain their thought processes, teachers will be able to discern the level of understanding. The vocabulary students use (or do not use) and the confidence with which they are able to answer probing questions can also provide insight into their levels of comprehension.

One of the most important features of formative assessment is that it actively involves students in their own learning. In assessment for learning, students are asked to reflect on their own work. They may be asked to consider multiple representations of a problem and then decide which of those representations makes the most sense, or which is the most efficient, or how they relate to one another. Students may be asked to make conjectures and then prove or disprove them by negation or counterexamples. Notice that it is the students doing the hard work of making decisions and thinking through the mathematical processes. Students who work at this level of mathematics, regardless of their grade level, demonstrate a deep understanding of mathematical concepts.

Introduction

Management and Assessment (cont.)

Assessment for learning makes learning a shared endeavor between teachers and students. In effective learning environments, students take responsibility for their learning and feel safe taking risks, and teachers have opportunities to gain a deeper understanding of what their students know and are able to do. Implementing a variety of tools and protocols when assessing for learning can help the process become seamless. Some specific formative assessment tools and protocols include:

- Student Response Forms or Journals
- Range Questions
- Round-Robin Activities
- Gallery Walks
- Observation Protocols
- Feedback
- Exit Cards

Student Response Forms or Journals

Providing students with an organized workspace for the problems they solve can help a teacher to better understand a student's thinking and more easily identify misconceptions. Students often think that recording an answer is enough. If students do include further details, they often only write enough to fill the limited space that might be provided on an activity sheet. To promote the expectation that students show all of their work and record more of their thinking, use the included *Student Response Form* (page 132; studentresponse.pdf), or have students use a designated journal or notebook for solving problems. The prompts on the *Student Response Form* and the additional space provided encourage students to offer more details.

Range Questions

Range questions allow for a variety of responses and teachers can use them to quickly gain access to students' understanding. Range questions are included in the activate section of many mini-lessons. The questions or problems that are posed are designed to provide insight into the spectrum of understanding that your students bring to the day's problems. For instance, you might ask *What do you know about the numbers 43, 53, 63?* Student responses may include stating that the factors of 63 are 3, 21, 9, and 7, or that 43 and 53 do not have factors other than 1 and the number itself, or that 43 and 53 are prime and 63 is composite. As you can imagine, the level of sophistication in the responses would vary and can help you decide which students to assign to which of the levels.

Round-Robin Activities

Round robin activities are designed to facilitate teachers' abilities to see in real time how proficient students are with various mathematical procedures and concepts. Students are grouped in triads and within each group students are assigned the number 1, 2, or 3. The teacher announces that all number 1s go to the board. The teacher dictates an expression, equation, or computation for these students to record and complete one step. After completing the step, the number 1s sits down and the number 2 students go to the board to do step 2. Upon completion of that step, these students return to their groups while the number 3s go to the board and do step 3. Students continue to go to the board in turn until the problem is complete. While the students are working at the board, the teacher has the opportunity to observe and provide individualized instruction and immediate feedback. At the same time, the format supports early intervention that prevents students from practicing, and thus reinforcing, any errors they make.

Management and Assessment (cont.)

Gallery Walks

Gallery walks can be used in many ways, but they all promote the sharing of students' problem-solving strategies and solutions. Pairs or small groups of students can record their pictures, tables, graphs, diagrams, computational procedures, and justifications on chart paper that they hang in designated areas of the classroom prior to the debriefing component of the lesson. Or, simply have students place their *Student Response Forms* at their workspaces and have students take a tour of their classmates' thinking. Though suggested occasionally for specific mini-lessons, you can include this strategy with any of the mini-lessons.

Observation Protocols

Observation protocols facilitate the data gathering that teachers must do as they document evidence of student learning. Assessment of learning is a key component in a teacher's ability to say, "I know that my students can apply these mathematical ideas because I have this evidence." Some important learning behaviors for teachers to focus on include: level of engagement in the problem/task; incorporation of multiple representations; inclusion of appropriate labels in pictures, tables, graphs, and solutions; use of accountable talk; inclusion of reflection on their work; and connections made between and among other mathematical ideas, previous problems, and their own life experiences. There is no one right form, nor could all of these areas be included on a form while leaving room for comments. Protocols should be flexible and allow teachers to identify categories of learning important to them and their students. An observation form is provided in the appendices (page 133; obs.pdf).

Feedback

Feedback is a critical component of formative assessment. Teachers who do not give letter grades on projects, quizzes, or tests, but who provide either neutral feedback or inquisitive feedback, find their students take a greater interest in the work they receive back than they did when their papers were graded. There are different types of feedback, but effective feedback focuses on the evidence in student work. Many students respond favorably to an "assessment sandwich." The first comment might be a positive comment or praise for something well done, followed by a critical question or request for further clarification, followed by another neutral or positive comment.

Exit Cards

Exit cards are an effective way of assessing students' thinking at the end of a lesson in preparation for future instruction. There are multiple ways in which exit cards can be used. A similar problem to the one students have previously solved can be posed, or students can be asked to identify topics of confusion, what they liked best, or what they think they learned from a lesson. Some teachers use them to inform the following day's instruction. If students show more misconceptions than understanding teachers use that information to add more practice with the concepts that caused the difficulty. Some exit card tasks are suggested in the *Differentiate* sections of the mini-lessons, but they may be added to any mini-lesson.

Introduction

How to Use This Book

Mini-Lesson Plan

Lessons are organized by **Common Core State Standards** domains.

Suggested **Problem-Solving Strategies** outline strategies students may want to use in solving the problem. However, these are not the only strategies that can be used to solve the problem.

The McREL mathematics **Standards** for each lesson are provided.

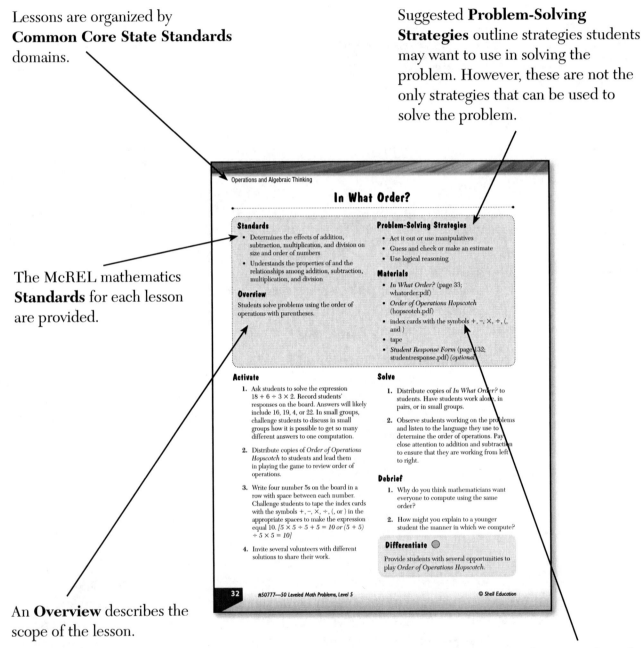

An **Overview** describes the scope of the lesson.

The **Materials** section lists the items needed for each lesson.

Introduction

How to Use This Book (cont.)

Mini-Lesson Plan (cont.)

The **Activate** section suggests how you can access or assess students' prior knowledge. This section might recommend ways to have students review vocabulary, recall experiences related to the problem contexts, remember relevant mathematical ideas, or solve simpler related problems.

The **Solve** section provides suggestions on how to group students for the problem they will solve. It also provides questions to ask, observations to make, or procedures to follow to guide students in their work.

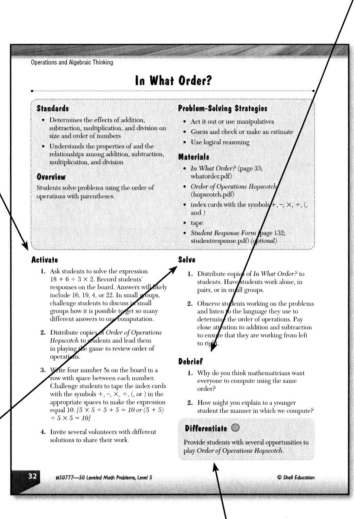

The **Debrief** section provides questions designed to deepen students' understanding of the mathematics and the problem-solving process. Because the leveled problems share common features, it is possible to debrief either with small groups or as a whole class.

The **Differentiate** section includes additional suggestions to meet the unique needs of students. This section may offer support for English language learners, scaffolding for below-level students, or enrichment opportunities for above-level students. The following symbols are used to indicate appropriate readiness levels for the differentiation:

- ●   below level
- ■   on level
- ▲   above level
- ★   English language learner

Introduction

How to Use This Book (cont.)

Lesson Resources

Leveled Problems

Each activity sheet offers **leveled problems** at three levels of challenge—below level, on level, and above level. Cut the activity sheet apart and distribute the appropriate problem to each student, or present all of the leveled problems on an activity sheet to every student.

Record-Keeping Chart

Use the **Record-Keeping Chart** (page 134) to keep track of the problems each student completes.

Observation Form

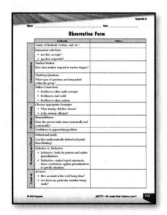

Use the **Observation Form** (page 133) to document students' progress as they work through problems on their own and with their peers.

Teacher Resource CD

Helpful reproducibles are provided on the accompanying **Teacher Resource CD**. A detailed listing of the CD contents can be found on pages 143–144. The CD includes:

- Resources to support the implementation of the mini-lessons
- Manipulative templates
- Reproducible PDFs of all leveled problems and assessment tools
- Correlations to standards

Introduction

How to Use This Book (cont.)

Lesson Resources (cont.)

Student Response Form

Students can attach their leveled problem to the form.

Students have space to show their work, provide their solution, and explain their thinking.

Appendix A

Name: _____ Date: _____

Student Response Form

Problem:

(glue your problem here)

My Work and Illustrations:
(picture, table, list, graph)

My Solution:

My Explanation:
_____
_____
_____

Introduction

Correlations to Standards

Shell Education is committed to producing educational materials that are research- and standards-based. In this effort, we have correlated all of our products to the academic standards of all 50 United States, the District of Columbia, the Department of Defense Dependent Schools, and all Canadian provinces. We have also correlated to the Common Core State Standards.

How To Find Standards Correlations

To print a customized correlation report of this product for your state, visit our website at **http://www.shelleducation.com** and follow the on-screen directions. If you require assistance in printing correlation reports, please contact Customer Service at 1-877-777-3450.

Purpose and Intent of Standards

Legislation mandates that all states adopt academic standards that identify the skills students will learn in kindergarten through grade twelve. Many states also have standards for Pre-K. This same legislation sets requirements to ensure the standards are detailed and comprehensive.

Standards are designed to focus instruction and guide adoption of curricula. Standards are statements that describe the criteria necessary for students to meet specific academic goals. They define the knowledge, skills, and content students should acquire at each level. Standards are also used to develop standardized tests to evaluate students' academic progress. Teachers are required to demonstrate how their lessons meet state standards. State standards are used in the development of all of our products, so educators can be assured they meet the academic requirements of each state.

McREL Compendium

We use the Mid-continent Research for Education and Learning (McREL) Compendium to create standards correlations. Each year, McREL analyzes state standards and revises the compendium. By following this procedure, McREL is able to produce a general compilation of national standards. Each lesson in this product is based on one or more McREL standards, which are listed in each lesson.

TESOL Standards

The lessons in this book promote English language development for English language learners. The standards listed on the Teacher Resource CD (tesol.pdf) support the language objectives presented throughout the lessons.

Common Core State Standards

The lessons in this book are aligned to the Common Core State Standards (CCSS). The standards listed on pages 27–31 (ccss.pdf) support the objectives presented throughout the lessons.

NCTM Standards

The lessons in this book are aligned to the National Council of Teachers of Mathematics (NCTM) standards. The standards listed on the Teacher Resource CD (nctm.pdf) support the objectives presented throughout the lessons.

Correlations to Standards (cont.)

Common Core State Standards Correlation

	Common Core Standard	Lesson
Operations and Algebraic Thinking	**5.OA.1** Use parentheses, brackets, or braces in numerical expressions, and evaluate expressions with these symbols.	In What Order?, page 32; Order Counts, page 34
	5.OA.2 Write simple expressions that record calculations with numbers, and interpret numerical expressions without evaluating them.	In What Order?, page 32; How Else Might I Look?, page 40
	5.OA.3 Generate two numerical patterns using two given rules. Identify apparent relationships between corresponding terms. Form ordered pairs consisting of corresponding terms from the two patterns, and graph the ordered pairs on a coordinate plane.	Geometric Patterns, page 38; Where Am I?, page 42; How Do I Change?, page 44; What's Our Relation?, page 46

Introduction

Correlations to Standards (cont.)

Common Core State Standards Correlation (cont.)

	Common Core Standard	Lesson
Numbers and Operations in Base Ten	**5.NBT.1** Recognize that in a multi-digit number, a digit in one place represents 10 times as much as it represents in the place to its right and $\frac{1}{10}$ of what it represents in the place to its left.	Name My Number, page 48
	5.NBT.5 Fluently multiply multi-digit whole numbers using the standard algorithm.	Rectangular Products, page 50; Grouping or Sharing?, page 54
	5.NBT.6 Find whole-number quotients of whole numbers with up to four-digit dividends and two-digit divisors, using strategies based on place value, the properties of operations, and/or the relationship between multiplication and division. Illustrate and explain the calculation by using equations, rectangular arrays, and/or area models.	Whatever Remains, page 52; Grouping or Sharing?, page 54
	5.NBT.7 Add, subtract, multiply, and divide decimals to hundredths, using concrete models or drawings and strategies based on place value, properties of operations, and/or the relationship between addition and subtraction; relate the strategy to a written method and explain the reasoning used.	Dealing with Decimals, page 56; Travel Expenses, page 60; Computing with Decimals, page 62; Dizzying Decimals, page 64; Estimating Decimals, page 66; About How Much?, page 68

Introduction

Correlations to Standards (cont.)

Common Core State Standards Correlation (cont.)

	Common Core Standard	Lesson
Number and Operations—Fractions	**5.NF.1** Add and subtract fractions with unlike denominators (including mixed numbers) by replacing given fractions with equivalent fractions in such a way as to produce an equivalent sum or difference of fractions with like denominators.	Fractional Sums, page 74; What's the Difference?, page 76; It's Close to What?, page 78; More or Less, page 80
	5.NF.2 Solve word problems involving addition and subtraction of fractions referring to the same whole, including cases of unlike denominators, e.g., by using visual fraction models or equations to represent the problem. Use benchmark fractions and number sense of fractions to estimate mentally and assess the reasonableness of answers.	Fractional Sums, page 74; What's the Difference?, page 76; It's Close to What?, page 78; More or Less, page 80
	5.NF.4 Apply and extend previous understandings of multiplication to multiply a fraction or whole number by a fraction.	The Product Is Smaller, page 84; Fair Sharing, Equal Groups, page 86; Map Reading, page 88
	5.NF.6 Solve real-world problems involving multiplication of fractions and mixed numbers, e.g. by using visual fraction models or equations to represent the problem.	The Product Is Smaller, page 84; Fair Sharing, Equal Groups, page 86; Map Reading, page 88
	5.NF.7c Solve real-world problems involving division of unit fractions by non-zero whole numbers and division of whole numbers by unit fractions, e.g., by using visual fraction models and equations to represent the problem. For example, how much chocolate will each person get if 3 people share $\frac{1}{2}$ lb of chocolate equally? How many $\frac{1}{3}$-cup servings are in 2 cups of raisins?	Fractional Areas, page 82

Correlations to Standards (cont.)

Common Core State Standards Correlation (cont.)

<table>
<tr><th colspan="2">Common Core Standard</th><th>Lesson</th></tr>
<tr><td rowspan="5">Measurement and Data</td><td>**5MD.1** Convert among different-sized standard measurement units within a given measurement system (e.g., convert 5 cm to 0.05 m), and use these conversions in solving multi-step, real-world problems.</td><td>What's My Unit?, page 98; Metrically Speaking, page 100; How Much Is There?, page 102</td></tr>
<tr><td>**5.MD.2** Make a line plot to display a data set of measurements in fractions of a unit ($\frac{1}{2}, \frac{1}{4}, \frac{1}{4}$). Use operations on fractions for this grade to solve problems involving information presented in line plots.</td><td>The Plot Thickens, page 112</td></tr>
<tr><td>**5.MD.3** Recognize volume as an attribute of solid figures and understand concepts of volume measurement.</td><td>Volume in Practice, page 96</td></tr>
<tr><td>**5.MD.4** Measure volumes by counting unit cubes, using cubic cm, cubic in, cubic ft, and improvised units.</td><td>Fill It Up, page 90; Cubic Views, page 94</td></tr>
<tr><td>**5.MD.5** Relate volume to the operations of multiplication and addition and solve real world and mathematical problems involving volume.</td><td>Volume in Practice, page 96</td></tr>
</table>

Correlations to Standards (cont.)

Common Core State Standards Correlation (cont.)

	Common Core Standard	Lesson
Geometry	**5.G.1** Use a pair of perpendicular number lines, called axes, to define a coordinate system, with the intersection of the lines (the origin) arranged to coincide with the 0 on each line and a given point in the plane located by using an ordered pair of numbers, called its coordinates. Understand that the first number indicates how far to travel from the origin in the direction of one axis, and the second number indicates how far to travel in the direction of the second axis, with the convention that the names of the two axes and the coordinates correspond (e.g., x-axis and x-coordinate, y-axis and y-coordinate).	Where Am I?, page 42; Plots A Lot, page 120; Graph It, page 130
	5.G.2 Represent real-world and mathematical problems by graphing points in the first quadrant of the coordinate plane, and interpret coordinate values of points in the context of the situation.	Where Am I?, page 42; Plots A Lot, page 120
	5.G.3 Understand that attributes belonging to a category of two-dimensional figures also belong to all subcategories of that category. For example, all rectangles have four right angles and squares are rectangles, so all squares have four right angles.	Congruency, page 116; Classifying Figures, page 118; Sort It Out, page 126
	5.G.4 Classify two-dimensional figures in a hierarchy based on properties.	Classifying Figures, page 118

Operations and Algebraic Thinking

In What Order?

Standards
- Determines the effects of addition, subtraction, multiplication, and division on size and order of numbers
- Understands the properties of and the relationships among addition, subtraction, multiplication, and division

Overview
Students solve problems using the order of operations with parentheses.

Problem-Solving Strategies
- Act it out or use manipulatives
- Guess and check or make an estimate
- Use logical reasoning

Materials
- *In What Order?* (page 33; whatorder.pdf)
- *Order of Operations Hopscotch* (hopscotch.pdf)
- index cards with the symbols $+, -, \times, \div, ($, and $)$
- tape
- *Student Response Form* (page 132; studentresponse.pdf) *(optional)*

Activate
1. Ask students to solve the expression $18 + 6 \div 3 \times 2$. Record students' responses on the board. Answers will likely include 16, 19, 4, or 22. In small groups, challenge students to discuss how it is possible to get so many different answers to one computation.
2. Distribute copies of *Order of Operations Hopscotch* to students and lead them in playing the game to review order of operations.
3. Write four number 5s on the board in a row with space between each number. Challenge students to tape the index cards with the symbols $+, -, \times, \div, ($, or $)$ in the appropriate spaces to make the expression equal 10. *[$5 \times 5 \div 5 + 5 = 10$ or $(5 + 5) \div 5 \times 5 = 10$]*
4. Invite several volunteers with different solutions to share their work.

Solve
1. Distribute copies of *In What Order?* to students. Have students work alone, in pairs, or in small groups.
2. Observe students working on the problems and listen to the language they use to determine the order of operations. Pay close attention to addition and subtraction to ensure that they are working from left to right.

Debrief
1. Why do you think mathematicians want everyone to compute using the same order?
2. How might you explain to a younger student the manner in which we compute?

Differentiate
Provide students with several opportunities to play *Order of Operations Hopscotch*.

Operations and Algebraic Thinking

In What Order?

Jeanie had to simplify each expression for a quiz and was told to justify her answer by showing all her computations. She used the order of operations to evaluate each expression. Is she correct? If so, how do you know? If not, how should she have simplified the expressions?

a. $7 - 5 + 3 \times 2$
$2 + 3 \times 2$
5×2
10

b. $8 + 4 \div 2 \times 2$
$12 \div 4$
3

In What Order?

Jose made up some practice problems to get ready for his math test. How might he use the order of operations to evaluate each expression? Show each step.

a. $3 + (9 \div 3 \times 3) - 5 \times 2$

b. $10 \div 5 \times (2 + 20 \times 3) - 15$

In What Order?

Mandy found the following challenge in a puzzle book. Use the symbols ( ), +, −, or ÷ to make a true equation. How can Mandy make each equation true?

a. $4 \quad 4 \quad 4 \quad 4 = 36$

b. $4 \quad 4 \quad 4 \quad 4 = 48$

© Shell Education   #50777—50 Leveled Math Problems, Level 5   33

Operations and Algebraic Thinking

Order Counts

Standards
- Determines the effects of addition, subtraction, multiplication, and division on size and order of numbers
- Understands the properties of and the relationships among addition, subtraction, multiplication, and division
- Understands the basic concept of an equality relationship

Overview
These problems focus on finding equivalent expressions. Students use order of operations to simplify a given expression to an equivalent simpler form.

Problem-Solving Strategy
Count, compute, or write an equation

Materials
- *Order Counts* (page 35; ordercounts.pdf)
- *Practice Expressions* (expressions.pdf)
- *Student Response Form* (page 132; studentresponse.pdf) *(optional)*

Activate
1. Divide students into groups of three. Assign the number 1, 2, or 3 to each student. Invite the number 1s to the board. Dictate the first expression from *Practice Expressions* to all students, even those at their seats.

    - Student number 1 copies the expression and completes the first step and sits down.
    - Student number 2 goes to the board and completes the second step and sits down.
    - Student number 3 goes to the board and completes the third step and sits down.
    - Student number 1 goes to the board and completes the next step and sits down.

2. Continue the round-robin until the expression is simplified. Read out the answer.

3. Pose the next expression and repeat the process in step 1.

Solve
1. Distribute copies of *Order Counts* to students. Have students work alone, in pairs, or in small groups.

2. Have students compare their answers with a partner. If any pair have a different answer invite the students to prove their answer is correct.

Debrief
1. How did you decide whether to multiply or divide first?

2. How did you decide whether to add or subtract first?

Differentiate △

Challenge students to include exponents when you are dictating the expressions.

Operations and Algebraic Thinking

Order Counts

Zoe is working on her math homework. She notices that some of the questions have multiple operations. How might she write an equivalent expression for each of the expressions using only two operations?

a. $18 - 6 + 5 \times 2$

b. $28 \div 2 \times 2 - 10 + 3$

Order Counts

Jesus is tutoring his friend Joaquin in math. The boys are writing equivalent expressions. What equivalent expressions with only two operations might the boys write?

a. $9 + 3 \times (16 \div 4 \times 2) - 4$

b. $(10 - 5 + 4) \div 3 \times 2$

Order Counts

Jackson really likes to be challenged in his work. He made up the following expressions for his partner Josh to evaluate. What is an equivalent expression that Josh might write using only one operation?

a. $3^3 \div 9 \times 4 + 10 - 2$

b. $7 - 3 + 4 \times 5^2 + (10 - 3 \times 2)$

Operations and Algebraic Thinking

Number Patterns

Standard
Recognizes a wide variety of patterns and the rules that explain them

Overview
Students complete mathematical number patterns and determine how two patterns are related.

Problem-Solving Strategies
- Act it out or use manipulatives
- Organize information in a picture, list, table, graph, or diagram

Materials
- *Number Patterns* (page 37; numberpatterns.pdf)
- *Pattern Examples* (patternexamples.pdf)
- circular discs or small tiles
- graph paper (graphpaper.pdf) *(optional)*
- *Student Response Form* (page 132; studentresponse.pdf) *(optional)*

Activate
1. Display *Pattern Examples* for students.
2. Challenge students to determine each of the patterns. Suggest students model the pattern using the discs or tiles.
3. Invite a volunteer to share a solution.
4. Ask students if anyone found the answer in a different way. If so, invite that student to show how he or she modeled the pattern. Repeat this with as many different representations as students made.

Solve
1. Distribute copies of *Number Patterns* to students. Have students work alone, in pairs, or in small groups.
2. Invite several students to share their strategies and solutions. Discuss all solutions that are presented.

Debrief
1. How did you decide what the pattern was?
2. How did you decide what the expression would be?
3. What other strategies might you use to determine how to continue a pattern?

Differentiate
Even when number patterns or sequences are numeric, it is helpful to all students to model them using discs or small tiles. The manipulatives encourage mathematical modeling. Some students will benefit from graphing the patterns to visualize how the patterns change.

Operations and Algebraic Thinking

Number Patterns

Olivia is solving number puzzles for fun. Help her find the four missing numbers in each pattern. How does the second pattern relate to the first?

0, 7, 14, 21, 28, _____, _____, _____, _____

0, 14, 28, 42, 56, _____, _____, _____, _____

Number Patterns

Olivia made up the following two patterns. Assuming the patterns continue, what might the next four numbers be? Show or explain your thinking. How does the second pattern relate to the first?

9, 11, 13, 15, 17, _____, _____, _____, _____

27, 33, 39, 45, 51, _____, _____, _____, _____

Number Patterns

Olivia found some really challenging puzzles. Write the next three numbers in each pattern. How does the second puzzle relate to the first? Show or explain your thinking. Create two new patterns that relate to each other.

10, _____, 30, _____, _____, 60, _____, _____, 90

2, 4, _____, _____, _____, 12, _____, 16, _____

© Shell Education   #50777—50 Leveled Math Problems, Level 5   37

Operations and Algebraic Thinking

Geometric Patterns

Standards
- Recognizes a wide variety of patterns and the rules that explain them
- Knows that a variable is a letter or symbol that stands for one or more numbers

Overview
Students complete geometric patterns and evaluate mathematical aspects of them.

Problem-Solving Strategies
- Act it out or use manipulatives
- Organize information in a picture, list, table, graph, or diagram

Materials
- *Geometric Patterns* (page 39; geopatterns.pdf)
- circular discs or square-inch tiles
- *Student Response Form* (page 132; studentresponse.pdf) *(optional)*

Activate
1. Have students explain why they think certain numbers are described as square numbers or triangular numbers.
2. Display a pattern using discs or tiles illustrating the first three iterations of a pattern.
3. Ask students to model the pattern using manipulatives. Challenge them to build the next three iterations of the pattern.
4. Have a volunteer model their work for the class.
5. Invite any students who did it differently to show how they modeled the pattern. Repeat this with as many different representations as students made.

Solve
1. Distribute copies of *Geometric Patterns* to students. Have students work alone, in pairs, or in small groups.
2. As students work, ask clarifying questions such as *How did you decide to create this pattern?*
3. Invite student volunteers to share their solutions. Ask if anyone did it differently or got a different answer.

Debrief
1. How did you decide what the pattern was?
2. How did you decide what the expression would be?
3. What other strategies might you use to determine the growth patterns?

Differentiate
Allow students as much time as they need to use manipulatives to model patterns. Students themselves should determine when they are ready to work without manipulatives.

Operations and Algebraic Thinking

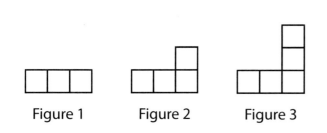

Connor built the first three figures using tiles. Look at the pattern he made. Then, complete the steps below.

1. Draw Figure 4.
2. Find the number of squares in Figure 10.
3. Write an expression or rule that would help you find the total number of squares in any figure.

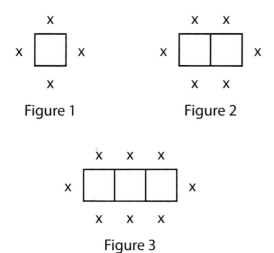

Rajon drew the three pictures. Look at the pattern he made. Complete the steps below.

1. Draw Figure 4.
2. Find the number of Xs in Figure 20.
3. Write an expression or rule that would help you find the total number of Xs for any figure.

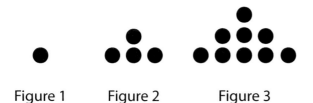

Isaiah made the following design using pennies. Help him:

1. Draw Figure 4.
2. Find the number of dots in Figure 100.
3. Write an expression or rule that would help you find the total number of dots in any figure.

Operations and Algebraic Thinking

How Else Might I Look?

Standards
- Determines the effects of addition, subtraction, multiplication, and division on size and order of numbers
- Understands the properties of and the relationships among addition, subtraction, multiplication, and division
- Understands the basic concept of an equality relationship

Problem-Solving Strategy
Count, compute, or write an equation

Materials
- *How Else Might I Look?* (page 41; mightlook.pdf)
- *Expression Practice* (expressionprac.pdf)
- *Student Response Form* (page 132; studentresponse.pdf) *(optional)*

Overview
These problems focus on equivalent expressions. Students examine representations of expressions and, without doing the mathematics, determine whether they are equivalent.

Activate
1. Distribute *Expression Practice* to students.
2. Present an expression and challenge students to write an equivalent expression.
3. Invite volunteers to share their solutions. Repeat this with as many different representations as students made.

Solve
1. Distribute copies of *How Else Might I Look?* to students. Have students work alone, in pairs, or in small groups.
2. Suggest that students evaluate each expression.
3. Invite a student volunteer to share a solution. Ask if anyone did it differently or if anyone got a different answer. Discuss all solutions that are presented.

Debrief
1. How might you check to ensure the expressions are equivalent?
2. Which of the expressions model the distributive property?

Differentiate ○ □ △ ☆
This is an appropriate time to use the mathematical vocabulary for the *commutative*, *associative*, and *distributive* properties. Have students make a personal glossary in which they identify the term, show an example of the term, record a strategy or mnemonic they can use to help them remember the term, and include what is important about the term.

Operations and Algebraic Thinking

Noah is working on his math homework. He noticed that all the expressions had the same numbers. Help Noah match up the two expressions that are equivalent.

$4 \times 8 + 3 \times 8$ $3 + 4 \times 8$ $(3 + 4 \times 8) \times 8$ $(4 + 3) \times 8$

Help Shay write equivalent expressions for the expressions below.

$15 \times 6 + 5 \times 15$

$12 \times (67 + 23)$

$(19 + 7) \times (4 + 7)$

Heather is rewriting her computations to show different equivalent expressions. She says the following expressions are equivalent. Is she correct? If so, why? If not, what should she have written?

$3 \times 78 + 4 = 4 \times 78 + 3$

$5 \times (67 + 15) = 5 \times 67 + 5 \times 15$

Operations and Algebraic Thinking

Where Am I?

Standard
Knows basic characteristics and features of the rectangular coordinate system

Overview
Students represent points on a coordinate plane and find distances from one point to another.

Problem-Solving Strategies
- Act it out or use manipulatives
- Organize information in a picture, list, table, graph, or diagram

Materials
- *Where Am I?* (page 43; whereami.pdf)
- chalk or masking tape
- graph paper (graphpaper.pdf)
- *Student Response Form* (page 132; studentresponse.pdf) *(optional)*

Activate
1. Before class begins, use chalk or masking tape to create a large coordinate plane in an outdoor space.
2. Have students make a list of everything they know about graphing.
3. If students do not mention them, introduce and discuss the terms *origin* (0, 0), *quadrant I* (+, +), *quadrant II* (−, +), *quadrant III* (−, −), and *quadrant IV* (+, −).
4. Distribute graph paper to students. Have them draw and label a coordinate plane on the paper.
5. Have students bring their graph paper and a pencil to the outdoor coordinate plane.
6. Invite a volunteer to stand at the origin. Then, ask the student to go to point (5, 3). Have the class plot the point on their grids. Repeat with several other points, discussing any differences in students' work.

Solve
1. Distribute copies of *Where Am I?* Instruct students to work on the problems individually, in pairs, or in small groups.
2. As students work, observe whether they are plotting the ordered pairs at the correct points.
3. Invite a student volunteer to share a solution. Ask if anyone got a different answer or did it differently.

Debrief
1. How did you decide which direction to move in first?
2. Does it matter if you graph the *y*-value before the *x*-value?

Differentiate

For students who display difficulty plotting the points in the problems, provide them with additional time modeling the process on the large outdoor grid. Physically moving to a point on the graph may help students internalize the order of the ordered pairs.

Operations and Algebraic Thinking

Where Am I? ○

Cole is learning to graph on the coordinate grid. Where would he place the following ordered pairs? Label each point A, B, C, D, or E.

A (−5, 3)   B (−4, −7)   C (6, −3)   D (1, 6)   E (−2, 0)

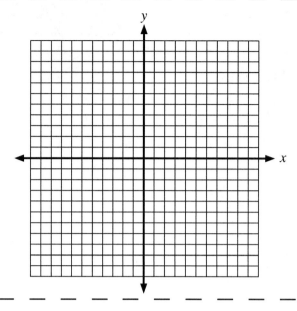

Where Am I? □

Jamal and some friends are going to walk from school to the soccer field after school. Their school is located at (−7, 4) while the soccer field is located at (5, −2). If each square on the coordinate plane represents one block and they can only walk on the sidewalks (gridlines), how far is the soccer field from the school? Justify your answer using the coordinate plane.

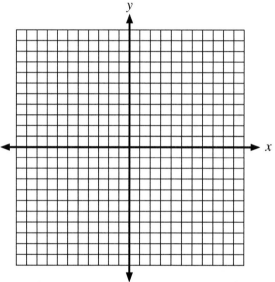

Where Am I? △

Isabelle lives at (5, 7) and is walking to her friend Abigail's house at (2, −8). Both girls are going to walk from there to the library, which is located at (−7, 8).

- Locate and label Isabelle's house.
- Locate and label Abigail's house.
- If each square represents one block and they can only walk on the sidewalks (gridlines), how far will Isabelle walk?

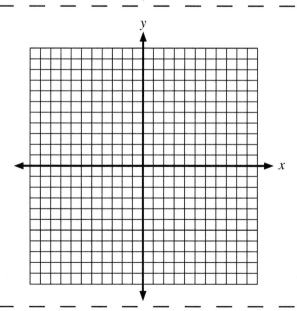

© Shell Education   #50777—50 Leveled Math Problems, Level 5   **43**

Operations and Algebraic Thinking

How Do I Change?

Standards
- Recognizes a wide variety of patterns and the rules that explain them
- Knows basic characteristics and features of the rectangular coordinate system

Overview
Students are given *x* and *y* values, find and express the relationship between the values, and use that relationship to fill in missing values.

Problem-Solving Strategies
- Find information in a picture, list, table, graph, or diagram
- Organize information in a picture, list, table, graph, or diagram

Materials
- *How Do I Change?* (page 45; change.pdf)
- graph paper (graphpaper.pdf)
- *Student Response Form* (page 132; studentresponse.pdf) *(optional)*

Activate
1. Display the following function table for students. Have students explain what they know about the table.

x	1	2	3	4	5	6
y	4	7	10	13	16	19

2. Ask students to describe the pattern with the *x* and *y* values *(x values are increasing by 1 and y values are increasing by 3)*.

3. Ask students what they think the rule is for the function table. Show students how to write the rule as an equation with variables *(y = 3x + 1)*.

4. Show students how to form ordered pairs from the function table and graph them on the coordinate grid.

Solve
1. Distribute copies of *How Do I Change?* to students. Have students work alone, in pairs, or in small groups.

2. Observe the strategies students use to find the functions.

3. Invite several volunteers to share their thinking.

Debrief
1. How did you decide what the rules were?

2. How might you use the relationship between the *x* and *y* values to find the rule?

Differentiate ○
Guide students to determine the relationship between how the *y* values changes in relation to how the *x* values changes.

44   #50777—50 Leveled Math Problems, Level 5   © Shell Education

Operations and Algebraic Thinking

How Do I Change? ○

x	y
1	3
2	5
3	7
4	9
5	11

Mrs. Garcia's students made function tables to share with a partner. Look at the table Ben made. Write the rule he was thinking about. Graph the ordered pairs on a coordinate grid.

Rule: _____

How Do I Change? □

x	y
1	5
2	
3	13
4	17
5	

Graham's partner made this function table. Help Graham complete the table and find the rule his partner was thinking about. Graph the ordered pairs on a coordinate grid.

Rule: _____

How Do I Change? △

x	y
1	1
2	4
	9
4	
	25

Justin made this function table. Fill in the missing values. What is his rule? Graph the ordered pairs on a coordinate grid.

Rule: _____

Operations and Algebraic Thinking

What's Our Relation?

Standards
- Recognizes a wide variety of patterns and the rules that explain them
- Understands that the same pattern can be represented in different ways

Overview
These problems focus on numerical and graphical patterns and the relationship between two or more patterns.

Problem-Solving Strategy
Organize information in a picture, list, table, graph, or diagram

Materials
- *What's Our Relation?* (page 47; relation.pdf)
- *Student Response Form* (page 132; studentresponse.pdf) *(optional)*

Activate
1. Have students explain how they might determine whether multiple mathematical expressions are related using numbers and graphs. Have students exchange ideas or conjectures with a partner.
2. Pose the following problem: *I have a machine that takes any number that is put into it, triples that number, and subtracts five.*
3. Tell students that you input 10 into your machine. Ask them to identify the output. (25)
4. Invite a student to share a solution. Ask if anyone found a different output.
5. Ask the class to think about a way in which they can take your original rule and double it. Have volunteers share their solutions.

Solve
1. Distribute copies of *What's Our Relation?* to students. Have students work alone, in pairs, or in small groups.
2. As students are working, note who is using appropriate mathematical vocabulary. If the mathematical language is missing, restate what students are saying with the appropriate mathematical vocabulary.
3. Observe student engagement with the task. If you notice that a student is not participating, pose a question to engage that student.

Debrief
1. Explain how the table or graph helped you find the missing values.
2. Explain how the table helped you write the rule as an equation.

Differentiate ▲
Challenge students to graph the values from the function tables. Have them write mathematical expressions for each problem and explain how the equation relates to the graph.

Operations and Algebraic Thinking

Louie has to fill in the table below for his homework. What are the missing values? Use the rule to write an equation to find any *x* value.

Rule: Multiply by 4, add 6

x	1	4	8	11	15
y					

Equation: _____

Avery is working on puzzles in her math class. She noticed that the two tables have something in common. Find the missing values. Write the rule for each function table as an equation.

x	3	7	9	14	21	n
y	21		63		147	

x	3	7	9	14	21	n
y	42		126			

Rule: _____

Rule: _____

Jessa and Renee are walking for charity. Jessa says she is going to charge sponsors $1.50 for every mile she walks. Renee says that is too much so she is going to charge her sponsors $3.00 up front then $0.75 per mile. If they walk 10 miles, which student will raise the most money? What rule describes Jessa's plan? Renee's plan?

Number and Operations in Base Ten

Name My Number

Standards
- Uses a variety of strategies in the problem-solving process
- Understands the basic meaning of place value

Overview
These problems focus on place value and expanded form.

Problem-Solving Strategies
- Organize information in a picture, list, table, graph, or diagram
- Use logical reasoning

Materials
- *Name My Number* (page 49; namenumber.pdf)
- *Place Value Number* (placevalue.pdf)
- place value charts *(optional)*
- *Student Response Form* (page 132; studentresponse.pdf) *(optional)*

Activate

1. Distribute *Place Value Number* to students and ask students to determine the number they see. Have students write the digits under the squares to name the number represented for each place value.

2. Have students discuss with a partner what number is represented in *Place Value Number*.

3. Ask several volunteers to share their solutions. Discuss any differences in responses.

Solve

1. Distribute copies of *Name My Number* to students. Have students work alone or in pairs.

2. Ask students to share their solutions. Highlight any different solution paths that resulted in the correct answer.

Debrief

1. What do you notice about the values in the list?
2. How did you determine the number?
3. Is there a different strategy that can be used?

Differentiate

Some students may need to work with a place value chart longer than other students. Also, some students will benefit from a challenge. In this instance you might have students work with base four or five. Working in base four means that instead of regrouping every ten units, the regrouping happens after four units and the place value names are powers of four. In base five the regrouping happens after five units and the place value names are powers of five. Working in another base reinforces the understanding of base ten and provides a challenge to those who need it.

Number and Operations in Base Ten

Name My Number

Tommy wrote a seven digit number. Use the place value clues and chart below to determine his number.

2 × 1,000,000   5 × 1   1 × 10,000

4 × 100   6 × 1,000   3 × 10

Millions	Hundred Thousands	Ten Thousands	Thousands	Hundreds	Tens	Ones

Name My Number

Sarah wrote a number puzzle with place value clues. What is Sarah's number?

9 × 100,000

3 × 10

7 × 100,000,000

2 × 10,000

8 × 1,000,000

4 × 1

6 × 100

5 × 1,000

1 × 10,000,000

Name My Number

Jill made a number puzzle. What is her number?

The place value of the 4 is 1,000 times the place value of the 9.

The place value of the 3 is 100,000 times the place value of the 2.

The place value of the 7 is $\frac{1}{100}$ the place value of the 9.

The place value of the 8 is 10,000,000 times the place value of the 6.

The place value of the 1 is 100 times the place value of the 5.

The place value of the 2 is $\frac{1}{10,000}$ times the place value of the 9.

The place value of the 5 is $\frac{1}{10}$ times the place value of the 7.

The place value of the 6 is $\frac{1}{1,000,000}$ times the place value of the 3.

Number and Operations in Base Ten

Rectangular Products

Standards
- Multiplies whole numbers
- Solves word problems and real-world problems involving number operations, including those that specify units
- Understands strategies for the multiplication of whole numbers

Overview
Students explore multiplication both mathematically and visually, finding the product of two factors and creating an array model to show the operation.

Problem-Solving Strategy
Organize information in a picture, list, table, graph, or diagram

Materials
- *Rectangular Products* (page 51; rectproducts.pdf)
- *Array Model* (arraymodel.pdf)
- graph paper (graphpaper.pdf)
- *Student Response Form* (page 132; studentresponse.pdf) *(optional)*

Activate
1. Ask students to think about the results of multiplying any two factors. Record students' responses. Students may say that multiplying two factors results in a bigger number, a product, or an answer. If no one mentions that the product of two numbers is visually represented by a rectangle, record it on the list.

2. Distribute graph paper to students and ask them to draw as many different diagrams as they can to represent the product of 6 × 15. Encourage students to think about decomposing the factors to make different looking rectangles. For example, (6 × 5) + (6 × 10) or (3 × 15) + (3 × 15) as well as (6 × 15) and (15 × 6), which are the same shape with a different orientation.

3. Invite students to share their diagrams. After students have shared their models, display *Array Model* to illustrate the array model for the distributive property. Discuss the representations and their meaning.

Solve
1. Distribute copies of *Rectangular Products* to students. Have students work alone, in pairs, or in small groups.

2. As students work, listen for accountable talk. If necessary, ask students how they determine which factor they are decomposing.

3. Invite several students to share their solutions. Discuss any misconceptions.

Debrief
1. What strategies did you use to decompose the factors?

2. How did you determine the final product?

Differentiate
Encourage students to draw the problem using pictures or on graph paper. Drawing out the problem helps students see that multiplication is repeated addition.

Number and Operations in Base Ten

Rectangular Products ○

There are 23 students in Orlando's class. Each student has earned 40 stamps for excellence in mathematics. How many total stamps did the class earn? Show your answer with a drawing and with numbers.

Rectangular Products □

On Monday, Angel bought 8 CDs that cost $15 each. On Saturday, he bought another 8 CDs that cost $12 each. How much money did Angel spend on CDs? Show your answer with a drawing and with numbers.

Rectangular Products △

Milai measured the garden to be 45 feet by 120 feet. She wants to find the area of the garden but doesn't have a pencil or paper to help her multiply.

How might Milai make the problem simpler to solve?

What is the area?

Draw a picture to support your reasoning.

Number and Operations in Base Ten

Whatever Remains

Standards
- Divides whole numbers
- Solves word problems and real-world problems involving number operations, including those that specify units

Overview
Students perform whole-number division to solve real-world problems and evaluate how to represent the remainder in a division problem.

Problem-Solving Strategies
- Act it out or use manipulatives
- Count, compute, or write an equation
- Organize information in a picture, list, table, graph, or diagram

Materials
- *Whatever Remains* (page 53; remains.pdf)
- packets of stickers
- *Student Response Form* (page 132; studentresponse.pdf) *(optional)*

Activate
1. Ask students what they know about the remainder in a division problem. Record their responses on the board.
2. Show the class several packets of stickers. Distribute one sheet of stickers to each student. Be sure to have some leftover.
3. Ask students what you might do with the leftover stickers. Raise the options of breaking up the sheets into smaller segments or saving the extras for another time if students do not.

Solve
1. Distribute copies of *Whatever Remains* to students. Have students work alone, in pairs, or in small groups.
2. As students work, listen for accountable talk. If necessary, ask students how they can determine if the remainder is too large.
3. Invite a student volunteer to share a solution. Ask if anyone did it differently or got a different answer. Allow other students to share their work and discuss any misconceptions.

Debrief
1. What strategies did you use to estimate your quotient before actually dividing?
2. What did you notice about the remainder in relation to the divisor?
3. How did you determine whether to drop the remainder, round up the quotient, or write the remainder as a fraction or decimal?

Differentiate ○ □ △ ☆
Consider assigning an exit-card task, such as the following: *What helps you decide what to do with a remainder when dividing?*

52 #50777—50 Leveled Math Problems, Level 5 © Shell Education

Number and Operations in Base Ten

Whatever Remains ○

Juan Carlos is ordering vans to transport the soccer team to a regional tournament. If each van holds 8 people, how many vans should he order to transport the 35 team members and 5 adults?

Whatever Remains □

Ling Su is sharing $25 with his three brothers. How much money will each boy get? Will there be any money leftover?

Whatever Remains △

Cameron is organizing his baseball card collection into a binder. Each sleeve in the binder holds 12 baseball cards. Cameron has 367 baseball cards and 21 sleeves. Does he have enough sleeves to hold all his cards? If not, how many more does he need?

Number and Operations in Base Ten

Grouping or Sharing?

Standards
- Uses a variety of strategies in the problem-solving process
- Multiplies and divides whole numbers

Overview
These problems focus on the different types of division problems (measurement and partitive) and how they relate to multiplication.

Problem-Solving Strategies
- Act it out or use manipulatives
- Count, compute, or write an equation
- Guess and check or make an estimate
- Organize information in a picture, list, table, graph, or diagram

Materials
- *Grouping or Sharing?* (page 55; grouping.pdf)
- graph paper (graphpaper.pdf)
- colored pencils (*optional*)
- multiplication charts (*optional*)
- *Student Response Form* (page 132; studentresponse.pdf) (*optional*)

Activate
1. Have students explain two reasons why we use division. Answers should include "fair share" (measurement) and to make equal groups (partitive).
2. Distribute graph paper and colored pencils (if desired) to students. Instruct them to shade in an array to represent 36 square units and to label the dimensions.
3. Invite volunteers to share their arrays and dimensions until all possible arrays and dimensions are identified.
4. Discuss the fact that the length and width of the array are factors, one is the divisor and the other is the quotient.

Solve
1. Distribute copies of *Grouping or Sharing?* to students. Have students work alone, in pairs, or in small groups.
2. As students are working, walk around and listen for accountable talk. Are students using correct terminology (*divisor, quotient, factor,* and *multiple*)?

Debrief
1. What strategies might you use to estimate the answer or check the reasonableness of your solution?
2. If you represented the division problem in fraction form, which value would you put in the numerator and which would you put in the denominator?
3. Were the problems you were doing fair sharing or grouping?

Differentiate
Provide a multiplication chart to students. Encourage students to estimate how many times the divisor evenly divides into the dividend in order to check the reasonableness of their answer.

Number and Operations in Base Ten

Grouping or Sharing? ○

Your school is collecting toiletries for troops overseas. You have to evenly divide a variety of toiletries. If you have 3,220 items and you put 12 items in each bag, how many soldiers will receive a package? Show your work in a picture or table and with numbers.

Grouping or Sharing? ☐

Help Eli write a division word problem using the values below.

| 1,311 | 57 | 23 |

Grouping or Sharing? △

Brad wants to write a division problem that results in the greatest quotient. He is going to use the digits 5, 6, 7, and 9 for the dividend and the digits 4 and 6 for the divisor. What division problem will give him the greatest answer? Write a story problem that might use these numbers.

Number and Operations in Base Ten

Dealing with Decimals

Standards
- Uses a variety of strategies in the problem-solving process
- Understands the basic meaning of place value
- Adds decimals

Overview
These problems focus on addition and subtraction with decimals.

Problem-Solving Strategies
- Guess and check or make an estimate
- Organize information in a picture, list, table, graph, or diagram

Materials
- *Dealing with Decimals* (page 57; dealdecimals.pdf)
- *Decimal Grids* (decimalgrids.pdf)
- colored pencils (*optional*)
- *Student Response Form* (page 132; studentresponse.pdf) (*optional*)

Activate
1. Have students explain everything they know about the decimal point and what it represents. Record their responses on the board.
2. Distribute copies of *Decimal Grids* and colored pencils (if desired) to students. Discuss with students that the value of one large square is equal to one. Have students identify the value of 10 small squares (*0.1*) and 1 small square (*0.01*).
3. Instruct students to shade in the area represented by 0.4 (*40 squares*). Invite volunteers to share the area.
4. Repeat with various other decimal values.

Solve
1. Distribute copies of *Dealing with Decimals* to students. Have students work alone, in pairs, or in small groups.
2. As you observe students, pose questions such as *Will your product be greater than or less than one whole?*

Debrief
1. How did you work with the decimal point in addition and subtraction?
2. How can you tell if you placed the decimal point in the correct location?

Differentiate ○ ◻ △ ☆
It is helpful for some students to work with one-by-ten grids to model tenths, and show sums with tenths before moving onto addition with hundredths or thousandths. Consider assigning the following exit-card task: *Write the sum for each of the following expressions: 0.4 + 0.07; 0.07 + 0.5; 0.09 + 0.13*

56   #50777—50 Leveled Math Problems, Level 5   © Shell Education

Number and Operations in Base Ten

If Tucker draws a model for the sum of 0.7 and 0.26 to show his little brother, what might it look like? Use the decimal grid to model the problem.

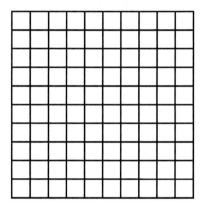

Gerardo found a quarter. He added that to the three dimes and seven nickels he already had in his pocket. How much money does Jose have now? Justify your answer with a picture or equation.

Myles asked his mother if he could use some of his birthday money. He explained that he wants to spend $6.78 of the $10.00 he has. How much money will Myles have after he makes his purchase? Justify your answer with a picture or equation.

Number and Operations in Base Ten

Expanded Form

Standards
- Understands the basic meaning of place value
- Understands the concepts related to decimals
- Uses base-ten concepts to represent decimals in flexible ways

Overview
These problems focus on expanded form of whole numbers and decimals.

Problem-Solving Strategy
Count, compute, or write an equation

Materials
- *Expanded Form* (page 59; expandedform.pdf)
- graph paper (graphpaper.pdf) *(optional)*
- *Student Response Form* (page 132; studentresponse.pdf) *(optional)*

Activate
1. Have students explain what they know about the number 4,569.271.
2. Model the meaning of expanding a number by asking your students to expand the number 45 using a power of ten. *(4 × 10 + 5 × 1)* From there, move on to expanding the number 123 *(1 × 100 + 2 × 10 + 3 × 1)*, followed by 3,456 *(3 × 1,000 + 4 × 100 + 5 × 10 + 6 × 1)*, followed by 567.89 *(5 × 100 + 6 × 10 + 7 × 1 + 8 × $\frac{1}{10}$ + 9 × $\frac{1}{100}$)*.
3. Discuss other possible representations of these numbers, for instance,
   $45 = 4 \times 10^1 + 5 \times 10^0$
   $123 = 1 \times 10^2 + 2 \times 10 + 3 \times 10^0$
   $3{,}456 = 3 \times 10^3 + 4 \times 10^2 + 5 \times 10^1 + 6 \times 10^0$
   $567.89 = 5 \times 10^2 + 6 \times 10 + 7 \times 10^0 + 8 \times 0.1 + 9 \times 0.01$ or
   $5 \times 10^2 + 6 \times 10^1 + 7 \times 10^0 + 8 \times 10^{-1} + 9 \times 10^{-2}$
4. Challenge your students to write the number 4,569.271 in expanded notation using place value notation.

Solve
1. Distribute copies of *Expanded Form* to students. Have students work alone, in pairs, or in small groups.
2. As students are working, ask them how they decide by which power of ten to multiply.
3. Ask students to explain what the expanded form represents.

Debrief
1. How did you decide what to multiply the digits to the right of the decimal point by?
2. What patterns did you notice?

Differentiate
Provide centimeter graph paper to students so they can visually represent the expanded form of numbers.

Number and Operations in Base Ten

Expanded Form ○

How might Katie write the numbers below in expanded form?

6,815

7.698

Expanded Form □

Emma wrote the following numbers in expanded form. What number did she write? Write her numbers in standard form.

$2 \times 10^4 + 5 \times 10^2 + 6$

$5 \times 10^3 + 1 \times 10^2 + 9 \times 10^0$

Expanded Form ◁

Josh wrote the number below in standard form. How might he write it using exponents? Write his number in expanded form.

75,397.063

Josh wrote the number below in expanded form. What is Josh's number? Write his number in standard form.

$4 \times 10^4 + 7 \times 10^2 + 3 \times 10^1 + 5 \times 10^{-1} + 2 \times 10^{-2}$

Number and Operations in Base Ten

Travel Expenses

Standard
Adds, subtracts, multiplies, and divides decimals

Overview
These problems focus on real-world situations that combine making change, calculating expenses with the division of whole numbers, and addition, subtraction, and multiplication of decimals.

Problem-Solving Strategies
- Act it out or use manipulatives
- Organize information in a picture, list, table, graph, or diagram

Materials
- *Travel Expenses* (page 61; expenses.pdf)
- items with price tags
- play money
- *Student Response Form* (page 132; studentresponse.pdf) *(optional)*

Activate
1. Describe a travel scenario to your students. Ask *If we were taking a trip to an amusement park for the whole day, what expenses do you think we would have to include in our budget?*
2. As students respond, list their answers on the board. Be sure to include gas, admission costs, lunch, snacks, and souvenirs.
3. Distribute play money to pairs of students. Have one student in each pair be the customer and the other be the cashier. The customer will select items to purchase, calculate the cost, and pay the cashier. The cashier will give back the appropriate change. Then have students switch roles.

Solve
1. Distribute copies of *Travel Expenses* to students. Have students work alone, in pairs, or in small groups.
2. As students work, ask clarifying questions such as *How do you know which math operation to use?*

Debrief
1. How do you decide what change to return to the customer?
2. How did you know what change was due when the customer gave you additional coins as well as a one-, five-, ten-, or twenty-dollar bill?
3. Did you find it easier to count on or subtract when calculating change? What made it easier?

Differentiate
Provide additional opportunities for students to model how much change they should receive from a cashier. Some students get very confused when the cost is say $0.84 and the customer gives $1.06. These students may benefit from working with you as their partner until they understand the concept of counting up with money.

60 #50777—50 Leveled Math Problems, Level 5 © Shell Education

Number and Operations in Base Ten

Travel Expenses

On a field trip from camp to the amusement park, Cam bought a drink and chips for $5.56 with tax. He gave the cashier a $20 bill. How much change did he receive?

Travel Expenses

Admission	
Adults (18 and over)	$26.50
Children (3–17)	$16.75
Children (1–2)	Free

Isabelle was in charge of buying the tickets for the amusement park. She has a brochure with the ticket prices displayed.

How much money will it cost to buy tickets if there are 12 adults traveling with the 51 fifth-grade students?

She paid the cashier with twelve one-hundred dollar bills and one quarter. How much change should she receive?

Travel Expenses

A summer camp staff took campers on a trip to an amusement park. Sixty-three people were going on the trip. The cost to rent vans that each hold 9–12 people was $138 per van. How many vans did the camp staff need to order? How much change was received if the rental fees were paid with $1,000 cash?

Number and Operations in Base Ten

Computing with Decimals

Standards
- Uses a variety of strategies in the problem-solving process
- Multiplies decimals

Overview
These problems focus on multiplication with decimals.

Problem-Solving Strategy
Organize information in a picture, list, table, graph, or diagram

Materials
- *Computing with Decimals* (page 63; computing.pdf)
- *Decimal Grids* (decimalgrids.pdf)
- *Student Response Form* (page 132; studentresponse.pdf) *(optional)*

Activate
1. Have students explain what they know about the decimal point.
2. Distribute copies of *Decimal Grids* to students. Instruct them to shade in the area formed by the factors 0.6 and 0.9. (*54 squares should be shaded to represent the product 0.54.*)
3. Invite a volunteer to share the array and its product. Ask if anyone did it differently until all possible arrays and dimensions are identified.

Solve
1. Distribute copies of *Computing with Decimals* to students. Have students work alone or in pairs.
2. As you observe students, pose appropriate questions, such as *Will your product be greater than or less than one whole?*
3. Invite volunteers to share their solutions. Ask if anyone did it differently or got a different answer. Invite those students to share their results, and discuss any misconceptions.

Debrief
1. What did you notice happening to the decimal point?
2. How can you tell if you placed the decimal point in the correct location?
3. Explain what happens to the decimal point when you multiply?

Differentiate
Using a visual representation such as the hundreds grid provides students with a visual representation of what happens when multiplying with decimals. Students should be encouraged to use the visual representations as long as they need them. Consider assigning the following exit-card task: *Find the product for each expression: 0.7×0.01; 0.07×0.1; and 0.007×0.01*

Number and Operations in Base Ten

Computing with Decimals ○

Brett is making a miniature model of his house and yard. The dimensions of the miniature will be 0.7 cm by 0.9 cm. What is the area of Brett's miniature model? Represent this on the grid and write an equation to solve the problem.

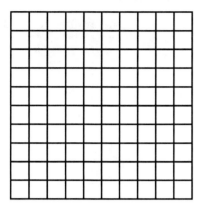

Computing with Decimals □

Lily is making a cardboard backing for her favorite photo. The photograph has a length of 5.75 inches and a width of 3.5 inches. What is the area of the cardboard backing that she will need?

Computing with Decimals △

Fill in the correct values to make a true statement. Each shape represents one value.

○ × ○ = 0.25

□ × ○ = 0.40

△ × □ = 0.56

○ = _____

□ = _____

△ = _____

© Shell Education   #50777—50 Leveled Math Problems, Level 5   **63**

Number and Operations in Base Ten

Dizzying Decimals

Standards
- Uses a variety of strategies in the problem-solving process
- Multiplies and divides decimals

Overview
Students divide decimal numbers to solve real-world problems.

Problem-Solving Strategies
- Organize information in a picture, list, table, graph, or diagram
- Work backward

Materials
- *Dizzying Decimals* (page 65; dizzying.pdf)
- *Decimal Grids* (decimalgrids.pdf)
- *Student Response Form* (page 132; studentresponse.pdf) *(optional)*

Activate
1. Have students consider whether dividing decimals results in a quotient that is greater than or less than the dividend.

2. Distribute copies of *Decimal Grids* to students and ask them to model the area of 0.27 on a grid. Have them identify 0.3 as the factor on one of the axes. Challenge them to identify the missing factor. *(0.9)* Ask whether the missing factor is greater than, less than, or equal to the area.

3. Invite volunteers to shade the grid. Ask if anyone did it differently. Challenge students to explain why the missing factor is greater than the dividend. *(If you take three-tenths of nine-tenths you are getting a smaller amount; if you take some of a number less than one whole, the amount you take keeps getting smaller.)*

4. Ask students to explain how successively multiplying 0.3 by multiples of 10 can help them find the quotient. *(If you start by multiplying by 10 the decimal moves to the right. All the other products are found by using the first answer. Since 0.3 × 10 = 3, 0.3 × 20 is double that product.)*

Solve
1. Distribute copies of *Dizzying Decimals* to students. Have students work alone, in pairs, or in small groups.

2. As students work, ask them to explain their thinking as they solve the problems.

3. Invite several volunteers to share their solutions and solution paths. Discuss any misconceptions.

Debrief
1. What strategies did you use to find the quotient?

2. How might you make a multiplication table to help you determine the quotients?

Differentiate

Show students how to model partial quotients by determining how the decimal divisor evenly divides the complete dividend. For example, ask students to determine how many three-tenths are in forty-three and five tenths, rather than by using traditional division algorithms.

Number and Operations in Base Ten

Dizzying Decimals ○

Jared bought 12 bags of chips for his party. If he paid $7.20 total, how much did each bag of chips cost?

Dizzying Decimals □

Each bottle of lemonade costs the local soccer team $1.75. They have $437.25 in their treasury to buy as many bottles of lemonade as possible to sell for a profit at their soccer games. How many bottles can the team afford to buy?

Dizzying Decimals △

Katie's grandfather gives each of his grandchildren the same amount of money for their birthdays. One year, he gave each grandchild $25.60 for a total amount of $716.80. How many grandchildren does he have?

Number and Operations in Base Ten

Estimating Decimals

Standards
- Divides decimals
- Uses specific strategies to estimate computations and to check the reasonableness of computational results

Overview
Students use front-end estimation to estimate quotients, then divide decimal numbers by whole numbers to find exact answers.

Problem-Solving Strategies
- Guess and check or make an estimate
- Organize information in a picture, list, table, graph, or diagram

Materials
- *Estimating Decimals* (page 67; estimatingdec.pdf)
- *Student Response Form* (page 132; studentresponse.pdf) *(optional)*

Activate

1. Have students estimate the quotient for 375.78 ÷ 34 and explain their reasoning.

2. Ask students to brainstorm situations when it is necessary to get an exact answer for a division problem involving decimals.

3. Have students rewrite the following expressions using front-end estimation: 342 ÷ 63 *(300 ÷ 60)*; 817 ÷ 32 *(800 ÷ 30)*; 463.48 ÷ 54 *(500 ÷ 50)*; 539.28 ÷ 67 *(500 ÷ 70)*

4. Have students estimate the quantities using the new expressions, then ask volunteers to share their estimates. Call attention to the place value to which students identify the estimated values. Ask if anyone got a different estimate. Share different strategies and solutions. Discuss any misconceptions.

Solve

1. Distribute copies of *Estimating Decimals* to students. Have students work alone, in pairs, or in small groups.

2. As students work, observe those students who estimate before computing and those who compute first then round their answers.

3. Ask several volunteers to share their solutions. Be sure to let students who did it differently show their work to ensure that there are no misconceptions.

Debrief

1. What is the advantage of using estimation to find quotients?

2. When do you think it is necessary to find an exact quotient?

Differentiate ○ □ △ ☆

Consider assigning an exit-card task such as the following: *Use front-end estimation to approximate 768 ÷ 43.*

Number and Operations in Base Ten

Dora is trying to estimate how many groups of 87 she can fit into 267.35. How might she use front-end estimation to get an approximate value?

A candy factory makes 293.75 pounds of candy every 16 minutes.

- Use front-end estimation to estimate the number of pounds per minute.
- Calculate exactly how many pounds of candy were made each minute.
- How close was your estimate to the exact number of pounds?

After winning the baseball tournament, Buddy's team of 19 players and their 2 coaches went out for pizza. The total cost of the pizza, drinks, and ice cream was $241.67. After a tip of $42.00 was added to the total, each person paid the same amount to cover the bill.

- Use front-end estimation to estimate the cost per person.
- Calculate exactly how much each person paid for the meal.
- How close was your estimate to the exact cost?

Number and Operations in Base Ten

About How Much?

Standards
- Uses a variety of strategies in the problem-solving process
- Uses specific strategies to estimate computations and to check the reasonableness of computational results

Overview
These problems focus on understanding and applying estimation strategies.

Problem-Solving Strategies
- Guess and check or make an estimate
- Use logical reasoning

Materials
- *About How Much?* (page 69; abouthowmuch.pdf)
- *Student Response Form* (page 132; studentresponse.pdf) *(optional)*

Activate

1. Display the following problem for students: *Members of the Collins Middle School choir were invited to sing to the president and his family. They will have to pay $1,278.50 for hotel rooms, $689.95 for meals, and $862.79 for transportation. If they include $700 for additional expenses, about how much will the trip cost?*

2. Have students estimate the answer and explain how they derived the estimation. Invite several students to share their strategies.

3. Display the following problem for students: *The art teacher orders very small cylinders in which she pours paint for her students. She buys 1,000 cylinders at a time and pays $25.75 for the containers. About how much does each cylinder cost?*

4. Ask students how they might estimate the cost of each container, and what a realistic estimate might be for the rough cost per container ($0.03). Ask several volunteers to share their estimates and solution strategies.

Solve

1. Distribute copies of *About How Much?* to students. Have students work alone, in pairs, or in small groups. Remind students to estimate the answers rather than find exact answers.

2. As students work, observe carefully to ensure that they are not adding or subtracting first then rounding their answers up or down. Have them explain how they are finding their estimates.

Debrief

1. How did you estimate the answer? Would any other estimate be reasonable?

2. What strategies did you use to find the estimates?

3. When is it advantageous to get an estimate for an answer rather than an exact amount?

Differentiate ○ □ △ ☆

Consider assigning an exit-card task, such as the following: *Explain in your own words how much smaller a millionth is compared with a thousandth. Give an example of something this weighs about a thousandth of a gram.*

Number and Operations in Base Ten

About How Much? ○

Sarah estimated that the sum of 58.47 + 123.058 is about 720. Is she correct? If not what might be a better estimate?

About How Much? □

Tim is adding the values below and wants to know if his answer is reasonable. Explain to him how he might estimate the sum to check the reasonableness of his answer.

4.09 + 84.163 + 9.607 + 478.2 = 989.61

About How Much? △

Kelly estimated the sum of 45.78 + 234.029 + 78.35 + 159.012 + 9.39 as 352. Her brother William estimated the same sum as 500. Whose estimate is closer? Justify your answer by showing how to use front-end estimation.

Number and Operations—Fractions

Where Do I Fit?

Standards
- Understands the concepts related to fractions
- Uses models to identify, order, and compare numbers

Overview
Students use a number line, picture, or diagram to order and compare fractions.

Problem-Solving Strategies
- Act it out or use manipulatives
- Use logical reasoning

Materials
- *Where Do I Fit?* (page 71; wherefit.pdf)
- sticky notes
- *Student Response Form* (page 132; studentresponse.pdf) *(optional)*

Activate
1. Ask students how they might decide which of two fractions is larger. Record their responses on the board.
2. Record the following fractions on sticky notes: $\frac{4}{7}$, $\frac{5}{8}$, $\frac{3}{4}$, $\frac{7}{8}$, and $\frac{9}{10}$. Sketch a large number line on the board with 0 and 1 marked.
3. Invite a volunteer to place one of the fractions on the number line. Ask the student to explain the placement of the fraction. Allow other students to share ideas and strategies on how to decide where the fraction should be placed.
4. Repeat step 3 with the remaining fractions, stopping to discuss the placement of each fraction as a class.

Solve
1. Distribute copies of *Where Do I Fit?* to students. Have students work alone, in pairs, or in small groups.
2. While students are working, pose questions such as *What strategies are you going to use to decide where to place each fraction?*
3. Identify students who appear to be guessing, relying on other students, or are disengaged. Engage those students in modeling fractions by folding congruent sheets of paper into the given fractions, then aligning them from smallest to greatest.

Debrief
1. How might you use benchmark fractions such as $\frac{1}{2}$ or $\frac{1}{4}$ to determine where a fraction belongs on the number line?
2. How do you know the fractions are in the correct place on the number line?

Differentiate ◐
Demonstrate various strategies for comparing fractions. Help students recognize that they can use common numerators as well as common denominators to compare fractions, and discuss how the numerator relates to the denominator.

Number and Operations—Fractions

Yolanda is trying to decide if she should choose $\frac{3}{5}$ or $\frac{5}{7}$ of her favorite dessert. She wants the larger of the two pieces. Which should she choose? Justify your solution using a number line.

Olivia and Zoe started with the same number of beads. Olivia used $\frac{5}{8}$ of her beads while Zoe used $\frac{4}{5}$ of her beads. Which of the girls has more beads left? Justify your solution using a number line. Support your solution with a written explanation of your answer.

Geri's homework assignment asked her to order some fractions from least to greatest. Arrange the fractions below on a number line. Support your solution with a written explanation of your answer.

$\frac{3}{8}$ $\qquad$ $\frac{5}{6}$ $\qquad$ $\frac{4}{5}$ $\qquad$ $\frac{5}{8}$ $\qquad$ $\frac{2}{3}$ $\qquad$ $\frac{1}{8}$

Number and Operations—Fractions

Ribbons and Bows

Standards
- Understands the concepts related to fractions
- Uses models to identify, order, and compare numbers

Overview
Students order and compare fractions and mixed numbers.

Problem-Solving Strategies
- Act it out or use manipulatives
- Count, compute, or write an equation
- Use logical reasoning

Materials
- *Ribbons and Bows* (page 73; ribbonsbows.pdf)
- sticky notes
- *Student Response Form* (page 132; studentresponse.pdf) *(optional)*

Activate
1. Have students explain how they know which is greater, $\frac{5}{8}$ of a pizza or $\frac{4}{7}$ of the same sized pizza?
2. Sketch a large number line on the board with the 0 and 1 identified.
3. Present four fractions to the class and challenge them to order the fractions from least to greatest.
4. Record the same fractions on sticky notes and invite a volunteer to the board to position the fractions on the number line.
5. Ask if any students did it differently. Invite those students to show where they located the fractions. Discuss the strategies students used.

Solve
1. Distribute copies of *Ribbons and Bows* to students. Have students work alone, in pairs, or in small groups.
2. While students are working, pose appropriate questions such as *What strategies are you going to use to decide which is the greatest length?*

3. As you observe students working, identify students who appear to be guessing, relying on other students, or are disengaged. Engage those students in modeling fractions by folding congruent pieces of paper into the given fractions, then aligning them from smallest to greatest.

Debrief
1. How can you use the numerator and denominator to determine if a fraction is greater than, less than, or equal to the given fraction?
2. Why don't we subtract the numerator from the denominator as a means of comparing fractions?
3. What other strategies might you use to compare fractions?

Differentiate ◯
Some students may need more time to work with fractions. Provide repeated support until they are able to compare and order fractions found on a ruler (*sixteenths, eighths, fourths, and halves*).

Number and Operations—Fractions

Ribbons and Bows

Dustin is tying a ribbon around a present. If he needs at least $\frac{5}{6}$ feet of ribbon to fit around the present, which length of ribbon should he choose: $\frac{1}{2}$ feet, $\frac{2}{3}$ feet, or $\frac{10}{12}$ feet?

Ribbons and Bows

Grace is organizing the bows she is selling for the Fourth of July. They measure $5\frac{1}{2}$ inches, $5\frac{1}{8}$ inches, $5\frac{1}{4}$ inches, $5\frac{3}{4}$ inches, and $5\frac{3}{8}$ inches. If she wants to display them from longest to shortest, in what order should she show them?

Ribbons and Bows

Abby is making bows for the school play. The size of the bows depends on the length of the ribbon she chooses. She must choose among ribbon that is $3\frac{7}{8}$ yards, $3\frac{13}{16}$ yards, $3\frac{5}{8}$ yards, $3\frac{5}{16}$ yards, $3\frac{1}{2}$ yards, $3\frac{9}{16}$ yards, $3\frac{1}{4}$ yards, and $3\frac{3}{4}$ yards long. She needs help ordering the ribbon from the shortest length to the longest. In what order should she place the ribbons?

Number and Operations—Fractions

Fractional Sums

Standards
- Adds simple fractions
- Solves word problems and real-world problems involving number operations, including those that specify units

Overview
These problems focus on adding and subtracting fractions.

Problem-Solving Strategy
Act it out or use manipulatives

Materials
- *Fractional Sums* (page 75; fractionalsums.pdf)
- blue, yellow, and green tiles
- *Fraction Strips* (fractionstrips.pdf)
- *Student Response Form* (page 132; studentresponse.pdf) *(optional)*

Activate
1. Have students explain what the numerator and the denominator represent in a fraction.
2. Distribute copies of *Fraction Strips* and colored tiles to pairs of students. Challenge students to model the sum of $\frac{2}{3}$ and $\frac{3}{4}$ using colored tiles. Suggest that they use the blue tiles for the first addend, the yellow tiles for the second addend, and the green tiles for the sum.
3. Remind students that adding fractions requires a common denominator. Review with students how to find the common denominator, as necessary.

Solve
1. Distribute copies of *Fractional Sums* to students. Have students work alone, in pairs, or in small groups.
2. As students work, listen to the mathematical language they use while discussing fractions. Are they saying *three-eighths* or *three over eight*?
3. Observe and note which students forget to record the common denominator and which students are proficient in using the appropriate vocabulary.

Debrief
1. What did you notice about the denominators when you added fraction values?
2. How does the numerator change when a common denominator is found?
3. What information does the denominator give?

Differentiate ○ ■ △ ★
Some students may benefit from modeling equivalent fractions on *Fraction Strips* before trying to add or subtract fractions. When using the strips, be sure to use factors of twelve for the denominators so the tiles fit on the strip. After students demonstrate proficiency modeling equivalent fractions, challenge them to add fractions with unlike denominators. Consider assigning an exit-card task such as the following: *Find the sum of $\frac{2}{3} + \frac{3}{4} + \frac{1}{5}$.*

Number and Operations—Fractions

Greyson is mixing red and blue paint to make purple paint. He puts $\frac{1}{4}$ cup of red paint in with $\frac{5}{8}$ cup of blue paint. How much purple paint did he make?

David spends $\frac{2}{3}$ hour playing video games, $\frac{1}{2}$ hour doing his chores, and $1\frac{1}{4}$ hours watching his brother. The rest of the time he can decide what to do. How much of David's time is already scheduled?

Samantha says that $\frac{2}{3}$ cup of cocoa plus $\frac{1}{2}$ cup of cocoa equals $\frac{3}{5}$ cup of cocoa. Andrew says that Samantha is incorrect, and that the sum is equal to $1\frac{1}{6}$ cups of cocoa. Who is correct? Justify your answer with pictures, numbers, or symbols.

Number and Operations—Fractions

What's the Difference?

Standard
Subtracts simple fractions

Overview
Students add and subtract fractions and mixed numbers in the context of real-world problems.

Problem-Solving Strategy
Organize information in a picture, list, table, graph, or diagram

Materials
- *What's the Difference?* (page 77; difference.pdf)
- *Student Response Form* (page 132; studentresponse.pdf) *(optional)*

Activate
1. Ask students to think about how they might explain to a third grader how to find the difference between two fractions or a fraction and a mixed number. Have students discuss this with a partner, and have several volunteers share their explanations with the class.
2. As a class, decide on the steps needed to subtract fractions and record them on the board.
3. Display the expression $3\frac{2}{5} - \frac{3}{4}$ on the board. Challenge students to use the directions recorded on the board to compute the difference.
4. Invite volunteers to share their solutions and record them on the board. Once all solutions are recorded, invite a student to the board to demonstrate how to compute the difference. Ask if anyone did it differently and invite those students to share their strategies. Be sure to include a pictorial representation as well as numeric solutions.

Solve
1. Distribute copies of *What's the Difference?* to students. Have students work alone, in pairs, or in small groups.
2. Assist students who struggle with rewriting the mixed number as an improper fraction by asking *How many of the denominator do you have?*

Debrief
1. Did you estimate what the difference might be? How did your estimate help you?
2. How might you describe a way to subtract a fraction from a mixed number?

Differentiate ⬤
Assist students in rewriting a whole number as an improper fraction to help them see how to subtract fractions from improper fractions with the same denominator. For example, $4 - \frac{3}{5}$ can be rewritten as $\frac{20}{5} - \frac{3}{5}$ or $3\frac{5}{5} - \frac{3}{5}$, or it can be modeled on a number line:

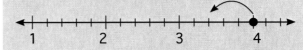

Number and Operations—Fractions

What's the Difference?

Lily is taking a quiz on fractions. She has to find the answers to the following four problems. Find the solutions. Justify your solution using numbers and a picture or diagram.

a. $\frac{3}{4} - \frac{3}{7}$

b. $1\frac{2}{3} - \frac{3}{5}$

c. $5 - \frac{3}{8}$

d. $4\frac{3}{5} - 3\frac{1}{2}$

What's the Difference?

Dante and some teammates celebrated a great game by eating large pizzas. Dante ate $\frac{2}{3}$ of a pizza, Leonardo ate $\frac{7}{8}$ of a pizza and Dakota ate $\frac{5}{6}$ of a pizza. If there were 6 pizzas purchased, how much pizza was left? Justify your solution using numbers and a picture or diagram.

What's the Difference?

Spencer is raking leaves to earn money for a new video game system. He figures he needs to work for a total of 25 hours to make enough money. He made the following schedule:

Day	Hours	Day	Hours
Monday	$1\frac{1}{2}$	Saturday	$4\frac{1}{4}$
Tuesday	$\frac{3}{4}$	Monday	$1\frac{2}{3}$
Wednesday	$2\frac{1}{4}$	Tuesday	$1\frac{2}{3}$
Thursday	$1\frac{3}{8}$	Wednesday	$1\frac{2}{3}$
Friday	$\frac{2}{3}$		

How many more hours will he need to work to have enough money? Justify your solution with words, pictures, and numbers.

Number and Operations—Fractions

It's Close to What?

Standards
- Understands the concepts related to fractions
- Adds and subtracts simple fractions

Overview
These problems focus on estimating the sum when adding fractions.

Problem-Solving Strategy
Organize information in a picture, list, table, graph, or diagram

Materials
- *It's Close to What?* (page 79; closetowhat.pdf)
- *Student Response Form* (page 132; studentresponse.pdf) *(optional)*

Activate
1. Have students explain how they might estimate the sum when adding fractions.
2. Sketch a number line on the board. Mark the following points: 0, $\frac{1}{2}$, and 1.
3. Challenge students to use the number line to estimate the sum of $\frac{1}{4}$ and $\frac{5}{8}$.
4. Emphasize that students do not have to place the fractions at an exact location on the number line, but should estimate the approximate placement of the fractions.

Solve
1. Distribute copies of *It's Close to What?* to students. Have students work alone, in pairs, or in small groups.
2. As students work, listen to the mathematical language they use with the fractions. Are they saying *three-eighths* or *three over eight*?
3. You may also observe which students are forgetting to record the common denominator and which students are proficient using the appropriate language.

Debrief
1. How did you decide whether the sum was greater than or less than $\frac{1}{2}$?
2. How did you decide where to place the fractions on the number line?
3. How did you determine your estimated sum?

Differentiate ○ ★
Some students may benefit from using a number line on which the benchmark fractions of 0, $\frac{1}{4}$, $\frac{1}{2}$, $\frac{3}{4}$, and 1 are already marked. They may need assistance in recognizing that it is helpful to compare the value of the numerator in relation to the denominator when determining whether the fraction is greater than, equal to, or less than one-half. For example, $\frac{3}{8}$ is less than $\frac{1}{2}$ because $\frac{4}{8}$ is equal to $\frac{1}{2}$ and $\frac{3}{8}$ is less than $\frac{4}{8}$.

Number and Operations—Fractions

Kiley used a number line to help decide where to place the sum of $\frac{3}{8} + \frac{3}{6}$. Where did Kiley place the sum? How do you know? Explain your answer.

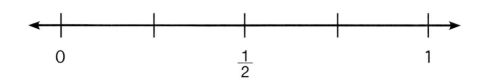

Connor says if he estimates the sum of $\frac{2}{4}$ and $\frac{2}{3}$, his sum will be close to but not quite 1. Is he correct? If so, explain how you know. If not, explain what a better estimate might be.

Anna Grace estimated that her brother Nick ate about $\frac{1}{3}$ of an apple pie, her brother Matt ate about $\frac{5}{8}$ of the pie, and she ate about $\frac{1}{4}$ of the pie. How might you convince Anna Grace that her estimates do not make sense? Use a picture or a diagram, words, and numbers to support your answer.

Number and Operations—Fractions

More or Less

Standards
- Understands the concepts related to fractions
- Adds and subtracts simple fractions

Overview
Students add and subtract mixed numbers in the context of real-world problems.

Problem-Solving Strategy
Organize information in a picture, list, table, graph, or diagram

Materials
- *More or Less* (page 81; moreless.pdf)
- *Number Line* (numberline.pdf)
- green, yellow, and blue colored tiles
- *Student Response Form* (page 132; studentresponse.pdf) *(optional)*

Activate
1. Have students explain how they might subtract fractions.
2. Distribute copies of *Number Line* and colored tiles to each student.
3. Ask students to find the difference between $\frac{5}{8}$ and $\frac{1}{4}$ using whichever method they choose.
4. If students choose to use the colored tiles, be sure they model the difference using the green tiles and the minuend with the blue or yellow tiles.
5. Have students share their answers.

Solve
1. Distribute copies of *More or Less* to students. Have students work alone, in pairs, or in small groups.
2. Listen to the mathematical language students use while discussing fractions. Are they saying *difference* or *common denominator*?
3. Note which students are checking their differences by writing a sum of the difference and the minuend.

Debrief
1. What did you notice about subtracting with fractions?
2. How did you use the colored tiles to model the difference problem?
3. How might you check your solution?

Differentiate
Help students understand that in finding the difference they are "undoing" addition. After students find the difference, challenge them to check their answer by modeling an addition problem using their answer and the minuend.

Number and Operations—Fractions

During one hour after school, LaDonna played basketball for $\frac{2}{8}$ of an hour. When she finished playing she spent $\frac{1}{3}$ of an hour doing her homework. What fraction of the hour does she have left to watch television or use the computer?

Henry lives $3\frac{5}{8}$ miles away from the Discovery Museum. Charlie lives $5\frac{1}{4}$ miles away from the same museum. How much farther away does Charlie live than Henry? Justify your answer with a picture, numbers, or words.

Marissa made money babysitting during summer vacation. Look at her schedule for one week:

Day	Hours
Monday	$3\frac{1}{2}$
Tuesday	$2\frac{5}{8}$
Wednesday	$5\frac{2}{3}$
Thursday	$3\frac{3}{4}$
Friday	$4\frac{3}{8}$

How many hours did Marissa work that week?

If Marissa had 60 waking hours during the week, how many hours did she have left after babysitting for other activities?

Number and Operations—Fractions

Fractional Areas

Standard
Understands the concepts related to fractions

Overview
Students identify fractional parts of a whole from drawings.

Problem-Solving Strategies
- Act it out or use manipulatives
- Use logical reasoning

Materials
- *Fractional Areas* (page 83; fractionalareas.pdf)
- construction paper
- scissors *(optional)*
- *Student Response Form* (page 132; studentresponse.pdf) *(optional)*

Activate
1. Have students explain what the denominator of a fraction represents.
2. Distribute construction paper to students. Direct students to find the area of the construction paper.
3. Instruct students to fold the construction paper in half diagonally and find the area of the triangle they made.
4. Discuss the relationship of the area of the triangle to the rectangle.
5. Fold the triangle in half again, and challenge students to find the area of the newly constructed triangle.
6. Discuss the relationship of this smaller triangle to the rectangle. How many of these triangles will fit into the original rectangle?

Solve
1. Distribute copies of *Fractional Areas* to students. Have students work alone, in pairs, or in small groups.
2. Invite student volunteers to share their responses. Encourage students with different solution paths to share their strategies.

Debrief
1. How did you decide which denominator to use?
2. What strategies did you use to determine the value of the fractional parts?

Differentiate

Provide scissors so that students can cut out the smallest figure in order to determine how many of them fit into the larger figure.

Number and Operations—Fractions

Mr. Letters is giving each of his children A, B, C, D, and E part of the land he owns in the country. How much land does each child receive?

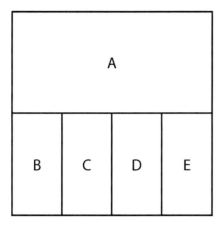

Ms. Rose is dividing her flower garden into different sections. She labeled them from A–J. What portion of her garden does each flower have?

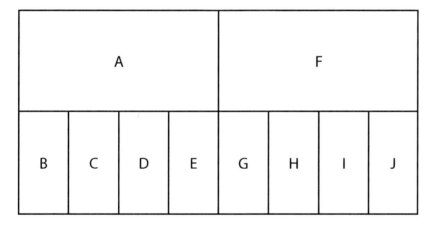

Jesse designed a new flag. He labeled each section of the flag with a letter. How much of the flag does each letter represent?

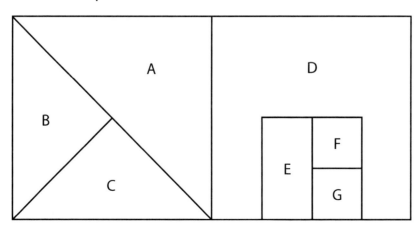

Number and Operations—Fractions

The Product Is Smaller

Standards
- Understands the concepts related to fractions
- Solves word problems and real-world problems involving number operations, including those that specify units

Overview
Students multiply fractions and draw an area model to support their solution.

Problem-Solving Strategy
Organize information in a picture, list, table, graph, or diagram

Materials
- *The Product Is Smaller* (page 85; productsmaller.pdf)
- colored pencils
- *Student Response Form* (page 132; studentresponse.pdf) *(optional)*

Activate
1. Ask students how they could visually represent two numbers multiplied together *(with an array or a rectangle)*.
2. Distribute several sheets of scrap paper and two colored pencils to each student. Explain that they will be folding and marking the paper to represent $\frac{2}{3} \times \frac{3}{4}$.
3. Have students hold their paper in landscape format and fold it into four rows to create fourths. Instruct students to shade in three of the rows to represent $\frac{3}{4}$.
4. Have students hold their paper vertically and fold it into three columns to create thirds. Instruct students to use a different color pencil to shade two of the columns to represent $\frac{2}{3}$.
5. Instruct students to open the folded paper and count the total number of rectangles *(12)*. Explain that this is the denominator. Then, direct them to count the number of rectangles shaded with both colors *(6)*. Explain that this is the numerator. So, the product of $\frac{2}{3}$ and $\frac{3}{4}$ is $\frac{6}{12}$.

Solve
1. Distribute copies of *The Product Is Smaller* to students. Have students work alone, in pairs, or in small groups.
2. Students may deduce that multiplying fractions can be done with "top times top and bottom times bottom." Rephrase those comments by stating *Do you mean finding the product of the numerators and the product of the denominators?*

Debrief
1. What did you notice about the size of the rectangle that is the product of two fractions?
2. What generalizations can you make about multiplying fractions?

Differentiate

Scaffold the instruction of this concept by guiding students through paper folding, then pictorial sketches of the area multiplication, followed by making a table of the fractions with which they compute. The table will show the factors and the products. Students should be guided to notice they are multiplying the numerators and the denominators.

84 #50777—50 Leveled Math Problems, Level 5 © Shell Education

The Product Is Smaller

David was asked to draw a picture of what multiplication looks like. He was told to show the product of $\frac{2}{3}$ and $\frac{4}{5}$. Show David's multiplication using an array. What is the product?

The Product Is Smaller

Destiny baked a pan of brownies. Her brother ate $\frac{1}{3}$ of the brownies. Destiny took $\frac{1}{2}$ of the remaining brownies to share with her friends. How much of the pan of brownies did she take? Draw a picture to support your solution.

The Product Is Smaller

Jill likes a morning snack. On Monday, she ate $\frac{1}{3}$ of the $2\frac{1}{2}$ pounds of strawberries she and her brother picked. How many pounds of strawberries did she eat? How many pounds were leftover? That afternoon, her brother ate $\frac{1}{4}$ of the strawberries that were left. Who ate more strawberries on Monday, Jill or her brother? How much more? Support your solution with a picture, numbers, or words.

Number and Operations—Fractions

Fair Sharing, Equal Groups

Standards
- Understands the concepts related to fractions
- Solves word problems and real-world problems involving number operations, including those that specify units

Overview
These problems focus on dividing fractions.

Problem-Solving Strategy
Organize information in a picture, list, table, graph, or diagram

Materials
- *Fair Sharing, Equal Groups* (page 87; fairequal.pdf)
- *Student Response Form* (page 132; studentresponse.pdf) *(optional)*

Activate
1. Ask students to think about the two reasons why division is used. Have students turn to a partner and discuss their ideas.
2. Ask groups to share what they discussed and record their theories on the board. If students do not include the idea that division is used for fair sharing and for making equivalent groups, share those reasons with the class and record them on the board.
3. Display the following problem for students: *Avery made a pan of brownies. She divided the pan into fourths and ate $\frac{1}{4}$. She is going to put $\frac{1}{8}$ of the remaining brownies into each sandwich bag. How many bags will she need?* Challenge students to model the division problem using either a picture or a number line.
4. Invite several students to share their models.

Solve
1. Distribute copies of *Fair Sharing, Equal Groups* to students. Have students work alone, in pairs, or in small groups.
2. As students work, listen to the mathematical language they use while discussing fractions. Are they saying *How many groups of*?

Debrief
1. What did you notice about how the quotient compared to the divisor?
2. Why do you think the answer in dividing a fraction by a fraction is greater than if you multiplied those same two fractions?

Differentiate ○ ◻ △ ☆
Consider assigning an exit-card task, such as the following: *Which gives the greater result $\frac{1}{3} \times \frac{4}{5}$ or $\frac{1}{3} \div \frac{4}{5}$?*

Number and Operations—Fractions

Fair Sharing, Equal Groups ○

Dakota has $\frac{7}{8}$ pound of candy. He is filling small bags with the candy. Each bag holds $\frac{1}{8}$ pound of candy. How many small bags of candy will Dakota make? Show how you know.

Fair Sharing, Equal Groups □

The science class is planting an urban garden in their schoolyard. They are going to divide $\frac{2}{3}$ of green space into smaller lots for each student. If each student is in charge of $\frac{1}{24}$ of that area, how many students are planning to garden? Draw a picture to support your solution.

Fair Sharing, Equal Groups △

Jermaine is splitting his penny collection into small stacks. If he has $3\frac{3}{4}$ pounds of pennies that he is going to package into $\frac{1}{2}$-pound rolls, how many rolls of pennies will he have? Justify your answer with a picture and numbers.

Number and Operations—Fractions

Map Reading

Standards
- Understands the concepts related to fractions
- Solves word problems and real-world problems involving number operations, including those that specify units
- Understands how scale in maps and drawings show relative size and distance

Overview
Students calculate distance in miles based on map legends that use fractions.

Problem-Solving Strategy
Organize information in a picture, list, table, graph, or diagram

Materials
- *Map Reading* (page 89; mapreading.pdf)
- *Maps* (maps.pdf)
- a variety of maps
- rulers
- *Student Response Form* (page 132; studentresponse.pdf) *(optional)*

Activate
1. Post a variety of maps around the classroom. Invite students to take a gallery walk visiting each map. At each map, instruct students to record what they notice about the map.
2. Discuss the terms *legend* and *inset*. Ask the students to discuss what they think the terms mean.
3. Distribute copies of *Maps* to students. Display the map and ask students to brainstorm and make a list of everything they know about the map.
4. Invite several students to share their results.
5. Challenge students to determine how many miles there are between each of the marked cities.
6. Invite students to compare their results with a partner. Have volunteers share their distances. Discuss any differences.

Solve
1. Distribute copies of *Map Reading* to students. Have students work alone, in pairs, or in small groups.
2. As students work, ask them how they are using the legend to find the miles.
3. Observe which students are simply guessing and which students are measuring and multiplying to find the distances.

Debrief
1. How did you use the legend to find the distances?
2. When using maps to determine distance, how exact do you think the solution must be?

Differentiate ○ ★
Assist students with reading a ruler. Maps with measures in inches require students to name and locate fractions on their rulers, and many students may need to learn how to read a ruler before they are able to successfully solve their problems.

Number and Operations—Fractions

Map Reading

A legend on a map showed the following:

$\frac{1}{4}$ inch = 12 miles

How many miles would $1\frac{1}{2}$ inches represent? Justify your answer using pictures and numbers.

Map Reading

Orion measured the distance between the towns of Pascal and Gauss to be $3\frac{1}{8}$ inches. The legend on the map showed $\frac{3}{4}$ inches = 8 miles. How many miles is it from Pascal to Gauss?

Map Reading

Luke knew that he traveled 33 miles from Euclid Middle School to his soccer match at Fibonacci Middle School. When he checked his road map, he noted that the legend was missing. He measured the distance between the two schools to be $2\frac{3}{4}$ inches. About how many miles are represented by a half-inch line? Justify your solution using a diagram or a picture and numbers.

Measurement and Data

Fill It Up

Standards
- Understands the basic measures of volume
- Selects and uses appropriate units of measurement, according to type and size of unit

Overview
Students find the volume of a cube or rectangular prism by counting cubes or multiplying the three dimensions.

Problem-Solving Strategies
- Act it out or use manipulatives
- Use logical reasoning

Materials
- *Fill It Up* (page 91; fillup.pdf)
- snap cubes or centimeter cubes
- graph paper (graphpaper.pdf)
- dot paper (dotpaper.pdf)
- *Student Response Form* (page 132; studentresponse.pdf) *(optional)*

Activate
1. Distribute cubes to pairs of students. Challenge them to make a cube that measures two units by two units by two units.
2. Direct students to record their dimensions and the total number of cubes (volume) they used. Use the terms *number of cubes* and *volume* interchangeably to help students understand the meaning of the term *volume*.
3. Increase the dimensions to two units by two units by four units. Have students sketch the rectangular prism on dot paper, build it, and record the dimensions and volume.
4. Challenge students to make as many different rectangular prisms as possible using 24 cubes. Invite volunteers to share their solutions. *(1 x 2 x 12; 2 x 2 x 6; 1 x 3 x 8; 1 x 4 x 6; 2 x 3 x 4; 1 x 1 x 24)*
5. Discuss which prism would use the least amount of cardboard to make a box. *(2 x 3 x 4; the closer to a perfect cube the less material needed to make the box.)*

Solve
1. Distribute copies of *Fill It Up* to students. Have students work alone, in pairs, or in small groups.
2. Make graph paper and cubes available for students who choose to use them.
3. Note whether students understand the difference between units of measure. They should use the term *cubic units* to discuss volume.

Debrief
1. What does volume represent?
2. How does the height of the object relate to the number of layers in the box?

Differentiate ▲
For a greater challenge, have students determine the dimensions of a prism given the volume.

Measurement and Data

Fill It Up

Sam has a box that is 6 inches by 7 inches by 8 inches. How many square-inch blocks will fit into Sam's box? Explain how you know using a picture and words.

Fill It Up

Jeremy has 72 square-inch cubes he plans to stack in a box. What are the different dimensions that the box might have? How do you know you have found them all?

Fill It Up

The Tasty Chocolate Company is looking for boxes to hold its new choco-mint cubes. They plan to package 36 pieces of 1 inch by 1 inch by 1 inch candies in each box. What are all the different sized boxes they might use? Which of the boxes is the most efficient if they measure efficiency by the least amount of cardboard they need to use to make the boxes?

Measurement and Data

How Spacious Is It?

Standards
- Understands the basic measures of perimeter and area
- Understands relationships between measures

Overview
These problems focus on area and perimeter of regular and irregular shapes.

Problem-Solving Strategy
Organize information in a picture, list, table, graph, or diagram

Materials
- *How Spacious Is It?* (page 93; spacious.pdf)
- *Dot Paper Polygons* (dotpolygons.pdf)
- rulers
- graph paper (graphpaper.pdf)
- *Student Response Form* (page 132; studentresponse.pdf) *(optional)*

Activate
1. Have students explain the difference between area and perimeter.
2. Display *Dot Paper Polygons* for students. Ask them to find the area of the figures.
3. Instruct students to use whatever strategies they want to determine the perimeter of the figures.
4. Ask students to share the strategies they used and then show how they found the areas and perimeters.

Solve
1. Distribute copies of *How Spacious Is It?* to students. Have students work alone, in pairs, or in small groups.
2. Provide graph paper and rulers so students can model the problems.

Debrief
1. How did you find the area?
2. How did you decide whether to use area or perimeter to find the largest lake?

Differentiate ○ □ △ ★
For students who have difficulty finding the area of the polygons on dot paper, have them work in pairs with students who are able to decompose the figures without difficulty. Consider assigning an exit-card task such as the following: *List three strategies you can use to find areas of various figures and shapes.*

Measurement and Data

Jaime is painting his rectangular bedroom walls blue. If the dimensions of the end wall are 8 feet by 10 feet, and the dimensions of the side wall are 10 feet by 10 feet, what is the area of his bedroom walls?

Find the area of the polygons below.

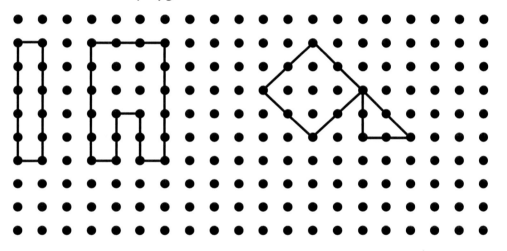

Walt and Sullivan both have fences around their yards. The area of both yards is the same, but the amount of fencing is different. If Walt and Sullivan's yards are each 300 square feet, how long could each of their fences be?

Measurement and Data

Cubic Views

Standards
- Understands the basic measures of volume
- Selects and uses appropriate units of measurement, according to type and size of unit

Overview
Students solve real-world problems to find the volume of a rectangular prism.

Problem-Solving Strategies
- Act it out or use manipulatives
- Organize information in a picture, list, table, graph, or diagram

Materials
- *Cubic Views* (page 95; cubicviews.pdf)
- dot paper (dotpaper.pdf)
- centimeter cubes
- *Student Response Form* (page 132; studentresponse.pdf) *(optional)*

Activate
1. Distribute dot paper and cubes to students.
2. Ask students to use centimeter cubes to build a rectangular prism that is 3 cm by 4 cm by 5 cm and draw the prism on dot paper. Have students determine the volume of the rectangular prism.
3. Lead students in developing the formula for finding the volume of a rectangular prism: $V = l \times w \times h$.

Solve
1. Distribute copies of *Cubic Views* to students. Have students work alone but allow them to compare solutions with a partner.
2. Invite several volunteers to share their solutions.

Debrief
1. How would you suggest a person go about finding the volume of a rectangular prism?
2. What does cubic units mean? How do you get them?
3. How can you use the number of cubes in the base to determine the total number of cubes in a prism?

Differentiate ▲
Have students determine how many centimeter cubes would fit in a cylinder that has a diameter of 7.5 cm and is 25 cm tall.

Measurement and Data

Cubic Views ○

Cole constructed a wooden box that had a base area of 72 square centimeters. How many layers of cubes will he need if he plans on putting 432 centimeter cubes in his wooden box? Justify your answer using numbers and a diagram.

Cubic Views □

Patrick drew a two-dimensional net for a three-dimensional prism. He cut it out to discover how many centimeter cubes will fit into it when he makes the box. He says the floor of the box holds 60 centimeter cubes. How many centimeter cubes will fit into the box when the sides are intact?

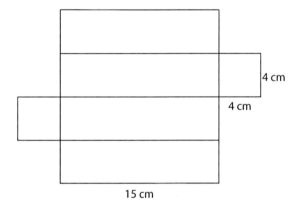

Cubic Views △

Manny built a rectangular prism with a length of 10 in., a width of 3 in., and a volume of 210 in.³ How high was the prism Manny built?

Measurement and Data

Volume in Practice

Standards
- Understands the basic measures of volume
- Selects and uses appropriate units of measurement, according to type and size of unit

Overview
Students apply their knowledge of finding volume to real-world scenarios.

Problem-Solving Strategies
- Count, compute, or write an equation
- Use logical reasoning

Materials
- *Volume in Practice* (page 97; volpracticc.pdf)
- *Student Response Form* (page 132; studentresponse.pdf) *(optional)*

Activate
1. Ask students how they might determine the total number of packages that can fit in a certain space.
2. Ask *Does it matter how the boxes are oriented?*
3. Ask students to share their responses.

Solve
1. Distribute copies of *Volume in Practice* to students. Have students work alone, in pairs, or in small groups.
2. Observe students at work. Pose clarifying questions, such as *Does the number of boxes that fit depend on how they are oriented? How might the boxes be stacked to take up the most space? How much space is used? How much space is leftover?*

Debrief
1. What are some considerations which need to be taken when packing a trailer truck?
2. What is an efficient way in which to determine how many boxes fit into a given space?

Differentiate ○ □ △ ☆
Consider assigning an exit-card task such as the following: *In your own words, describe how to determine the volume of a rectangular prism.*

Measurement and Data

Volume in Practice ○

Semi trailer trucks are rectangular prisms. The average height for the tractors is 9 feet. Most trailers are about 8 feet wide and 20 feet long. What is the volume that the tractor trailer can haul? Justify your answer with numbers and words.

Volume in Practice □

A shipping company wants to ship its boxes in a semi trailer truck. The boxes have dimensions of 2 feet by 5 feet by 3 feet. How many boxes will fit in a semi trailer truck that measures 9 feet by 8 feet by 20 feet? Justify your answer with numbers and a picture.

Volume in Practice △

The shipping company needs to stack boxes in the trailer so that the bottom of each box measures 4 feet by 6 feet and has a height of 3 feet. If the trailer is 10 feet high by 8 feet wide by 20 feet long, how much space is leftover?

Measurement and Data

What's My Unit?

Standard
Knows approximate size of basic standard units and relationships between them

Overview
Students determine correct units of measure for common items, or compare and order related units of measure.

Problem-Solving Strategies
- Act it out or use manipulatives
- Guess and check or make an estimate

Materials
- *What's My Unit?* (page 99; whatunit.pdf)
- variety of common objects such as paper clips, erasers, pens, pencils, markers, etc.
- rulers and balance scales
- *Student Response Form* (page 132; studentresponse.pdf) *(optional)*

Activate
1. Ask students to use their hands to show how long they think 10 centimeters is.
2. Distribute a variety of objects to small groups of students. Challenge students to predict the length and weight of each object.
3. Instruct students to make a table to record their estimates and actual measures.
4. Invite several volunteers to share their estimates and actual lengths and weights.

Solve
1. Distribute copies of *What's My Unit?* to students. Have students work alone or in pairs.
2. After students complete their estimates and measures, ask student volunteers to share their strategies and solutions.

Debrief
1. How did you decide which unit of measure to use?
2. How did you decide whether to use inches or centimeters?

Differentiate ●■▲★
Consider assigning an exit-card task, such as the following: *Draw a line that is approximately five and a half inches long.*

Measurement and Data

Julia needs to determine which measures are appropriate for each object. Help her match the objects to the most appropriate unit of measure.

| gallon | pint | quart |

Steve mixed up all the metric measures and is challenging you to put them in order from largest to smallest. Use pictures or words to explain how you know the correct order.

| 8 cm | 8 km | 8 dm | 8 m | 8 mm |

Cameron is helping his mother make labels. She wants to order the containers from least to greatest. How should he organize them? Justify your response by showing the equivalences between each unit of measure from least to greatest.

| quart | ounce | gallon | pint |

Measurement and Data

Metrically Speaking

Standard
Understands relationships between measures

Overview
Students convert metric units while solving real-world problems.

Problem-Solving Strategy
Organize information in a picture, list, table, graph, or diagram

Materials
- *Metrically Speaking* (page 101; metrically.pdf)
- *Metric Conversion* (conversion.pdf)
- metric conversion chart
- *Student Response Form* (page 132; studentresponse.pdf) *(optional)*

Activate
1. Ask students to predict which measure is larger, 63 meters or 63,000 mm.
2. Display *Metric Conversion* and challenge students to determine the equivalences.
3. Record the conjectures students make on the board. Instruct students to find a counterexample to each conjecture.

Solve
1. Distribute copies of *Metrically Speaking* to students. Have students work alone or in pairs.
2. Direct students to model the problems as they work through the conversions.
3. As you observe students at work, watch carefully to ensure they understand they are moving the decimal point as they move among the metric system and not just adding zeros.

Debrief
1. How did you decide which direction to move the decimal point? Why does it move?
2. How much larger is a kilometer than a millimeter? How do you know?

Differentiate ○ ■ △ ☆
Provide students with a metric conversion chart before moving on to the abstract understanding that they are multiplying and dividing by powers of ten and therefore moving the decimal point.

Measurement and Data

Metrically Speaking ○

Jared has a baseball bat that is one meter in length. How many millimeters long is his baseball bat? Justify your answer with a diagram, numbers, or words.

Metrically Speaking □

Sarah measured a dime in her wallet. She discovered the diameter is about 1 centimeter. How many dimes would Sarah need to line up to have a length of 5 meters? Justify your answer with a diagram, numbers, or words.

Metrically Speaking △

Isaac and his mom are running in a 10K race. Isaac usually runs in a 500-meter dash at school. How many more meters will Isaac run when he and his mom compete in the 10K? Justify your answer with a diagram, numbers, or words.

Measurement and Data

How Much Is There?

Standards
- Solves word problems and real-world problems involving number operations, including those that specify units
- Understands relationships between measures

Overview
Students use metric conversions to solve problems related to capacity.

Problem-Solving Strategy
Organize information in a picture, list, table, graph, or diagram

Materials
- *How Much Is There?* (page 103; howmuchthere.pdf)
- cylinders, liquid measuring cups, beakers, and test tubes labeled in metric units
- colored water
- *Student Response Form* (page 132; studentresponse.pdf) *(optional)*

Activate
1. Prepare stations around the classroom with liquid measuring cups, beakers, or test tubes with milliliters labeled on the side. Fill each to various heights with colored water. Number each station and label each container with a letter.
2. Ask students how they might determine the amount of colored water given the variety of cylinders filled to different levels. Demonstrate how to read the meniscus, or curve in the upper level of the liquid, to get an accurate measure.
3. Group students in triads and instruct them to record the height of the water at each station. Allot about 5–7 minutes at each station. Have students record their measurements on chart paper, indicating the station number and container letter.
4. After all the measurements are made instruct the students to post their charts around the room.
5. Invite the class to do a gallery walk to examine the measurements, looking for discrepancies, outliers, and similarities. Discuss students' observations.

Solve
1. Distribute copies of *How Much Is There?* to students. Have students work alone, in pairs, or in small groups.
2. Direct students to model the volumes as they work through the conversions.

Debrief
1. What is the unit measure when working with volume or capacity?
2. How did you determine how the given volumes compared with the unit of measure?

Differentiate ▲
Challenge students to determine whether a container with a larger diameter and shorter height hold more liquid than a container with a smaller diameter and taller height. Allow them time to explore this question using various containers.

Measurement and Data

How Much Is There?

Abby has 146 milliliters of water in her water bottle. Her sister Emma has 146 centiliters of water in her water bottle.

- Which of the girls has the most water?
- How much more does she have than her sister?
- Justify your answer with numbers, a diagram, or words.

How Much Is There?

Stephanie's class saves water in a rain barrel to water the school garden. The rain barrel holds 15,750 milliliters of water. Stephanie used 7.375 liters to water the tomato plants. How many liters of water is left in the water barrel?

How Much Is There?

Marisol has 15 liters of juice to share with her 23 classmates at the end of the year picnic. Jacinta said she wanted at least a half-liter of juice. If Marisol pours 24 equal glasses of juice, how many milliliters will each student receive? Did Jacinta receive at least a half-liter of juice? If so, how much more? If not, how much more juice should Jacinta get? Justify your answer with numbers, a diagram, or words.

Measurement and Data

All in a Line

Standards
- Understands that data represent specific pieces of information about real-world objects or activities
- Understands that spreading data out on a number line helps to see what the extremes are, where the data points pile up, and where the gaps are

Overview
Students represent data on a line plot and find the mean, median, mode, and range of a data set.

Problem-Solving Strategy
Organize information in a picture, list, table, graph, or diagram

Materials
- *All in a Line* (page 105; allinaline.pdf)
- easel-sized centimeter graph paper
- *Student Response Form* (page 132; studentresponse.pdf) *(optional)*

Activate
1. Display a list of 15 randomly-organized integers. Ask students what they know about the data. Record students' responses.
2. Have students make a line plot of the 15 numbers. Ask them what they can now say about the data. Record their responses.
3. If students do not mention the median, mode, range, or mean as attributes of the data, review these terms and their meanings.

Solve
1. Distribute copies of *All in a Line* to students. Have students work in pairs or in small groups. Distribute easel-sized graph paper to each group and instruct students to use it to make their line plot.
2. Direct students to use the line plot to determine the mode, range, median, and/or mean of the data.

Debrief
1. How did you use the data in the line plot to determine the mode of the data?
2. How did you use the data in the line plot to determine the range of the data?

Differentiate ◯
Students may benefit from using a pan balance or balance beam to model the concept of mean. If the average is correctly calculated, the pan or beam balance scales will be balanced. Seeing concretely what it means to average out a set of data is extremely beneficial for all students.

Measurement and Data

Members of the track team run after school every day and on Saturdays. At the end of each week the number of miles are posted in the locker room. Create a line plot that represents the data set.

| 20 | 14 | 26 | 18 | 21 | 19 | 14 | 23 | 26 | 18 | 14 | 17 | 18 | 23 | 23 |

Members of the track team run after school every day and on Saturdays. At the end of each week the number of miles are posted in the locker room.

| 20 | 14 | 26 | 18 | 21 | 19 | 14 | 23 | 26 | 18 | 14 | 17 | 18 | 23 | 23 |

Based on the data above, determine the median, mode, and mean.

Median _____

Mode _____

Mean _____

Members of the track team run after school every day and on Saturdays. At the end of each week the number of miles are posted in the locker room.

| 20 | 14 | 26 | 18 | 21 | 19 | 14 | 23 | 26 | 18 | 14 | 17 | 18 | 23 | 23 |

If all the miles run were evenly distributed among the runners, how many miles would each runner have?

Measurement and Data

What Is the Favorite?

Standards
- Understands that a summary of data should include where the middle is and how much spread there is around it
- Organizes and displays data in simple bar graphs, circle graphs, and line graphs

Overview
Students collect, organize, and display data in bar graphs and circle graphs, and find the mean, median, and mode of a data set.

Problem-Solving Strategies
- Find information in a picture, list, table, graph, or diagram
- Organize information in a picture, list, table, graph, or diagram

Materials
- *What Is the Favorite?* (page 107; favorite.pdf)
- *Student Response Form* (page 132; studentresponse.pdf) *(optional)*

Activate
1. Have students explain everything they know about how surveys are designed.
2. Ask students what kind of information they would like to have data about. Do they care about the flavors of ice cream their friends like, or the video games they play, or their favorite activities?
3. Discuss how they might collect, organize, and display the data. If students do not mention tally charts, circle graphs, and bar graphs, bring them up and ask what students know about them.
4. Model a sample survey by asking students whether their favorite lunch food is pizza, burgers, tacos, or other. Invite a volunteer to record students' responses using tallies.
5. Direct pairs of students to represent the data from the tally chart in a bar graph and a circle graph. Students will probably need guidance in converting percents to degrees in the circle graph. Engage students in a dialogue about how best to make the conversions.

Solve
1. Distribute copies of *What Is the Favorite?* to students. Have students work alone, in pairs, or in small groups.
2. As students are working, observe which students are leading the discussion and which ones rely on their partners.

Debrief
1. What do you need to consider when framing survey questions?
2. About how large should the "other" category be?

Differentiate ○ ☆
Some students may benefit from working with data that equals one-hundred subjects. This makes the calculations easier while still allowing for appropriate multiple representations of the data in circle graphs, bar graphs, and tallies.

Students represented the results of a survey they took about favorite sports in a circle graph. They interviewed a total of 200 students. How many students picked each sport? Justify your answers.

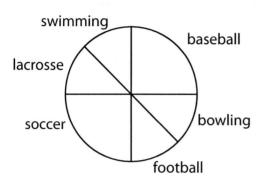

Sport	Students
baseball	
swimming	
bowling	
football	
soccer	
lacrosse	

The following data represent the number of sit-ups students in Ms. Kim's class completed during physical education tests.

$$57, 32, 98, 54, 76, 89, 87, 54, 51, 71, 37, 66, 90$$

- Create a bar graph that represents the data.
- Identify the mean.
- Interpret the data. What conclusions might you make about the physical fitness of students in Ms. Kim's class?

Fifth-grade students participated in a survey about how much time they spend using their cell phones each day during vacation week. The table below shows how students responded.

- Make a circle graph to represent the data in the table.
- What conclusions can you draw about the data? Be sure to include references to mean, mode, and median.

	1 hour	2 hours	3 hours	4 hours	5 hours
cell phone usage	|||	||||	𝍬 𝍬	||		

Measurement and Data

Stem-and-Leaf Plots

Standards
- Understands that a summary of data should include where the middle is and how much spread there is around it
- Understands that data come in many different forms and that collecting, organizing, and displaying data can be done in many ways

Overview
These problems focus on representing data in stem-and-leaf plots.

Problem-Solving Strategy
Organize information in a picture, list, table, graph, or diagram

Materials
- *Stem-and-Leaf Plots* (page 109; stemleaf.pdf)
- connecting cubes
- pan balances or balance beams
- *Student Response Form* (page 132; studentresponse.pdf) *(optional)*

Activate
1. Have students explain to you everything they can about organizing data.
2. Ask students to make a list of their classmates' birthdays (use only the day, for example, for August 10, use only 10). Have students put the data in order from least to greatest.
3. Show students how to create a stem-and-leaf plot using some of the data they collected. Have students complete the plot for the rest of the data set.
4. Invite several students to share their solutions.

Solve
1. Distribute copies of *Stem-and-Leaf Plots* to students. Have students work alone, in pairs, or in small groups.
2. As students are working on the problems, pose clarifying questions to help you uncover students' thinking.
3. Ask students to explain the difference between the mean and median and to give an example of when it is more important to use the median and when it is more important to use the mean.

Debrief
1. Explain how the stem-and-leaf plot uses place value.
2. How might you check the median in a stem-and-leaf plot?

Differentiate ○ □ △ ☆
Consider assigning an exit-card task such as the following: *Define the term* mean *or* average *in your own words. Give an example to support your answer.*

Measurement and Data

The data in the table was gathered to display a school's basketball scores over the last season. The coach has asked you to make a stem-and-leaf plot for this data. How might it look?

| 65 | 73 | 68 | 75 | 73 | 82 | 68 | 57 | 74 | 61 | 67 | 35 |

Mr. Wiley wants to know whether the mean, median, or mode will be the best measure to use to describe how well his students did on his math test. Which should he use and why?

```
 1 |
 2 | 5
 3 |
 4 | 0
 5 | 5 5
 6 | 0 5
 7 | 0 0 0 5 5 5
 8 | 0 0 0 5
 9 | 5 5
10 | 0
```

The Main Street School Eagles are competing against the Pumas to see which group can do the most squats in a two-minute period. They represented their data in a double stem-and-leaf plot.

Eagles	Stems	Pumas
	1	
	2	
5 4 4 3 3 2	3	0 1 2 4 4 8
8 8 8 7 6	4	1 2 2 6 7
6 4 3 1 1	5	1 3 3 3 4 5 6
5 4 3 2 1	6	0 0 1 4
	7	
	8	
1	9	

- What is the mode for the Eagles? Pumas?
- What is the median for the Eagles? Pumas?
- What can you say about the two teams based on the data?

© Shell Education #50777—50 Leveled Math Problems, Level 5 109

Measurement and Data

What Does It Mean?

Standards
- Understands that data represent specific pieces of information about real-world objects or activities
- Understands that a summary of data should include where the middle is and how much spread there is around it

Overview
Students find the mean and median of a data set.

Problem-Solving Strategy
Use logical reasoning

Materials
- *What Does It Mean?* (page 111; whatmean.pdf)
- connecting cubes
- pan balances or balance beams (*optional*)
- *Student Response Form* (page 132; studentresponse.pdf) (*optional*)

Activate
1. Have students explain what they know about the term *mean*.
2. Provide small groups of students with a pan balance or balance beam and a number of connecting cubes. Direct students to make six stacks of connecting cubes so that one has 3 cubes, one has 7 cubes, one has 10 cubes, one has 2 cubes, one has 6 cubes, and one has 5 cubes.
3. Next, ask students to rearrange the number of cubes so that each of the six stacks has the same number of connecting cubes. Ask students how many cubes are in each stack when the cubes are fairly distributed among the piles and how they know.
4. Invite a student volunteer to share their work. Ask if anyone did it differently. Ask if anyone got a different answer.

Solve
1. Distribute copies of *What Does It Mean?* to students. Have students work alone, in pairs, or in small groups.
2. As students are working on the problems, pose clarifying questions to help you uncover students' thinking.
3. Ask students to explain the difference between the *mean* and *median* and to give an example of when it is more important to use the median and when it is more important to use the mean.

Debrief
1. Explain how you determined the mean of a data set.
2. In what circumstances is the mean the measure of central tendency of choice?

Differentiate
Many students benefit from using manipulatives to build an understanding of mean and median. To find the median provide easel-sized grid paper, cut into strips. Students record the number of letters in each student's name in ascending order, one number per square. Invite students to fold the strip in half. The crease of the fold identifies the median name length.

Measurement and Data

What Does It Mean?

Mary and her five friends took a mathematics test. The scores on the tests were 65, 92, 87, 75, 80, and 100. If the total points are evenly distributed among the six students, how many points would each student earn?

What Does It Mean?

Charlie earned an 82, 75, 90, 65, and 70 on his spelling tests. Should he ask his teacher to report his median score or his average score to his parents? Justify your response.

What Does It Mean?

Jessa earned an 82, 95, 80, 75, 90, and 80 on her first six tests. What is the lowest score she must earn on her next test to ensure she has an average of at least 85? Justify your solution.

Measurement and Data

The Plot Thickens

Standard
Understands that spreading data out on a number line helps to see what the extremes are, where the data points pile up, and where the gaps are

Overview
Students redistribute fractions on a line plot so each fractional value is equivalent to the others, finding the mean of a data set.

Problem-Solving Strategies
- Find information in a picture, list, table, graph, or diagram
- Organize information in a picture, list, table, graph, or diagram
- Use logical reasoning

Materials
- *The Plot Thickens* (page 113; plotthickens.pdf)
- *Student Response Form* (page 132; studentresponse.pdf) *(optional)*

Activate

1. Display the following four fractions for students: $\frac{3}{4}$, $\frac{5}{8}$, $\frac{1}{2}$, and $\frac{7}{8}$.

2. Ask students to make four equivalent fractions. You may need to review how to create equivalent fractions.

3. Invite student volunteers to share their solutions. Ask if anyone did it differently or got a different answer.

Solve

1. Distribute copies of *The Plot Thickens* to students. Have students work alone, in pairs, or in small groups.

2. Model the use of academic vocabulary words such as *equivalent* and *common denominator*.

3. Ask clarifying and refocusing questions, such as *What measure of central tendency you are finding when all the values have the same amount? When all the fractions have the same amount, where on the line plot will the equivalent values be located? How do you know?*

4. Invite student volunteers to share their solutions. Ask if anyone got a different answer or did it differently.

Debrief

1. What did the original line plot tell you about the amount of liquid in each beaker?

2. What did the reorganized line plot tell you about the data?

Differentiate ▲
Have students who are ready for a greater challenge work with mixed numbers or decimals.

Measurement and Data

Jackson made a line plot to record the amount of liquid in a series of beakers. Now he wants to redistribute the liquid so each beaker holds the same amount. How much liquid will be in each beaker when he levels the amount in each of the nine beakers?

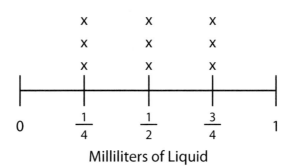

Billy made a line plot to record the amount of liquid in a series of beakers. Now he wants to redistribute the liquid so each beaker holds the same amount. How much liquid will be in each beaker when he levels the amount in each of the twelve beakers?

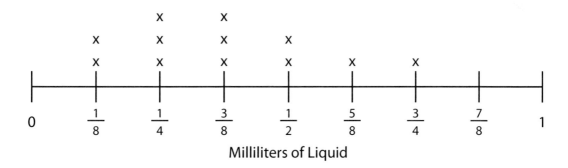

Logan made a line plot to record the amount of liquid in a series of beakers. Now he wants to redistribute the liquid so each beaker holds the same amount. How much liquid will be in each beaker when he levels the amount in each of the eighteen beakers?

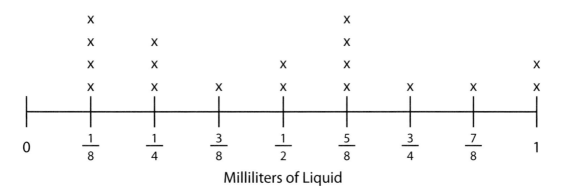

Measurement and Data

One or the Other

Standards
- Understands that the word "chance" refers to the likelihood of an event
- Recognizes events that are sure to happen, events that are sure not to happen, and events that may or may not happen
- Uses basic sample spaces to describe and predict events

Overview
Students solve problems involving binomial probability.

Problem-Solving Strategy
Use logical reasoning

Materials
- *One or the Other* (page 115; oneother.pdf)
- pennies or two-colored discs
- *Student Response Form* (page 132; studentresponse.pdf) *(optional)*

Activate
1. Have students explain what they think *binomial probability* means. Break down the term *binomial* into *bi* which means "two" and *nomial* which means "term."
2. Challenge students to think of situations where there are only two possible outcomes. Record all responses.
3. Ask students what the possible outcomes are when flipping pennies. Ask whether they are more likely to get heads or tails. Ask them to explain their reasoning.
4. Instruct pairs of students to toss a coin 20 times and record the results. Record the results for each pair on a class table. This will allow for a greater sample size.
5. Ask students to predict the probability of getting a heads on flip number 250. Explain that their prediction is considered the *expected value*.

Solve
1. Distribute copies of *One or the Other* to students. Have students work alone, in pairs, or in small groups.
2. As students are working, model academic vocabulary such as *likely*, *more likely*, and *less likely*.
3. Ask clarifying and refocusing questions such as *Why do you think the probability is greater than one-half? Do you think a table or organized list might help you find all the possible outcomes?*
4. Invite several volunteers to share their solutions.

Debrief
1. What do you think is the clearest way to organize the data?
2. How can you determine whether your probability outcome makes sense?

Differentiate
Provide students with a template for organizing their data.

Measurement and Data

One or the Other

Anne flipped a penny three times. She got two heads and one tail. What is the probability of her flipping the penny a fourth time and getting heads? Tails?

One or the Other

Jacob is flipping a two-colored disc. One side is yellow, the other side is red.
- What is the probability he will get a yellow?
- What is the probability he will get a red?
- What is the probability he will get a green?
- What is the probability that he will get a yellow or a red?

One or the Other

Katie has a bag filled with the numbers 1–10. She has 10 numbers altogether. She wrote a list of clues to describe the numbers in her bag. What numbers could be in her bag?
- It is more likely to get an even number than an odd number.
- It is impossible to get a 1.
- It is equally likely to get a prime number and a composite number.
- It is less likely to get a number less than 5 than a number greater than 5.

Geometry

Congruency

Standards
- Knows basic geometric language for describing and naming shapes
- Understands that shapes can be congruent or similar

Overview
Students decompose congruent figures.

Problem-Solving Strategies
- Act it out or use manipulatives
- Organize information in a picture, list, table, graph, or diagram

Materials
- *Congruency* (page 117; congruency.pdf)
- *Congruent Figures* (congruentfigs.pdf)
- construction paper
- scissors
- *Student Response Form* (page 132; studentresponse.pdf) *(optional)*

Activate
1. Have students make a list of everything they know about congruent figures.
2. Distribute copies of *Congruent Figures* to students.
3. Challenge students to sketch in nine congruent figures that have the exact same shape as the one on the paper.
4. Ask a student volunteer to share a solution. Ask if anyone did it differently or got a different answer.
5. Discuss the fact that the triangle, when decomposed into the nine smaller triangles, replicates Sierpinski's triangle as well as a beginning fractal.

Solve
1. Distribute copies of *Congruency* to students. Have students work alone, in pairs, or in small groups. Direct students to decompose each figure into four smaller congruent figures.
2. Provide students with construction paper and scissors should they choose to model the problems.
3. Invite student volunteers to share their solutions.
4. Ask if anyone got a different answer or did it differently.

Debrief
1. How did you decide the best way to sketch your lines to decompose the figures?
2. How did you check to ensure all the figures are congruent?

Differentiate ▲
Challenge students to make their own large figure that can be decomposed into at least four congruent shapes.

Geometry

How might Paige divide this rectangle into four smaller congruent triangles?

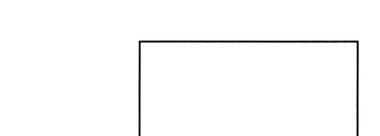

How might Jeb divide this triangle into four smaller congruent triangles?

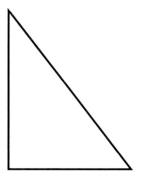

Sam knows that this trapezoid can be divided into four congruent trapezoids but he forgets how. Help Sam find a way to draw the congruent trapezoids.

Geometry

Classifying Figures

Standards
- Knows basic geometric language for describing and naming shapes
- Understands basic properties of figures

Overview
Students identify properties of polygons.

Problem-Solving Strategies
- Act it out or use manipulatives
- Organize information in a picture, list, table, graph, or diagram

Materials
- *Classifying Figures* (page 119; classifying.pdf)
- *Classifying Geometric Figures* (geofigures.pdf)
- scissors
- glue
- *Student Response Form* (page 132; studentresponse.pdf) *(optional)*

Activate
1. Have students explain what a polygon is.
2. Distribute copies of *Classifying Geometric Figures* to students. Have students organize the given shapes into the correct column.
3. Have students cut out the figures and glue them into the correct columns or sketch the figures in the correct columns.
4. Record all students' descriptions of what makes a polygon.

Solve
1. Distribute copies of *Classifying Figures* to students. Have students work alone or in pairs.
2. Ask students to explain their thinking and as they do, listen for correct mathematical vocabulary.
3. Invite a volunteer to share a solution. Ask if anyone did it differently or if they got a different answer. Invite those who did to share their solutions.

Debrief
1. How did you decide which figures shared the same attributes for a polygon?
2. What are some of the characteristics that all polygons have?
3. How might you classify the following figures: rhombus, square, parallelogram, rectangle, trapezoid, quadrilateral? Of these figures, which is the most generic?

Differentiate ○ ■ △ ☆
The vocabulary of geometric figures is often responsible for misconceptions. Encourage students to make their own dictionary of geometric terms, which should include examples of each geometric figure.

Geometry

Mary and Gail are working together to sort shapes. Which of the following shapes can be labeled polygons? Which are non-polygons?

Explain what you think the attributes of a polygon are.

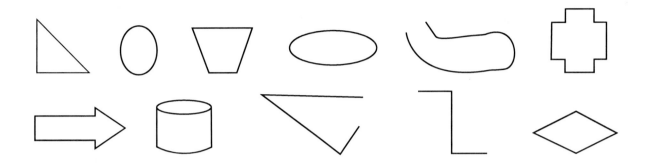

How might Frankie complete the Venn diagram using words that describe the attributes of rectangles and squares?

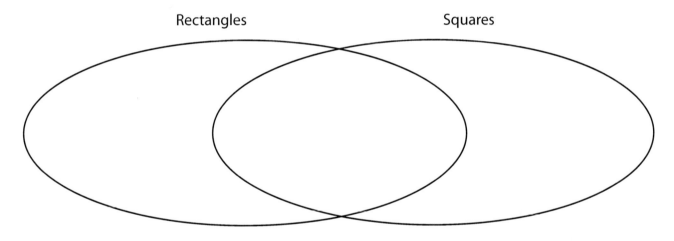

Help Matt complete the Venn diagram with words that describe the attributes of parallelograms and trapezoids.

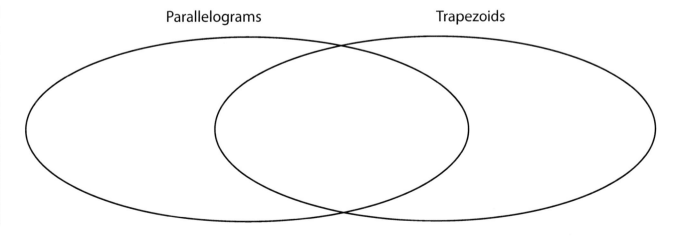

Geometry

Plots A Lot

Standards
- Knows basic geometric language for describing and naming shapes
- Understands basic properties of figures
- Knows basic characteristics and features of the rectangular coordinate system

Overview
Students plot points on the coordinate plane to make geometric figures.

Problem-Solving Strategy
Act it out or use manipulatives

Materials
- *Plots A Lot* (page 121; plotsalot.pdf)
- graph paper (graphpaper.pdf)
- rulers
- *Student Response Form* (page 132; studentresponse.pdf) *(optional)*

Activate
1. Distribute graph paper and rulers to students. Model drawing two intersecting, perpendicular lines (*x*- and *y*-axes). Have students draw this on their paper.
2. Guide students in labeling the axes, the origin, and the four quadrants.
3. Tell the students to use either + or − symbols within parenthesis (+, +), (−, +), (−, −), (+, −) to indicate the signs of numbers that are located in each quadrant.
4. Review plotting ordered pairs by dictating several for students to plot on their coordinate planes. Invite volunteers to share their responses with the class.

Solve
1. Distribute copies of *Plots A Lot* to students. Have students work alone, in pairs, or in small groups.
2. As students graph the points and complete the geometric figures, ask them to explain how they know in which quadrant to plot their points.
3. Invite several volunteers to share their solutions.

Debrief
1. What helped you decide how to complete each figure?
2. Did you think you could complete the figure in a different quadrant?

Differentiate
Allow students plenty of opportunities to practice plotting points in the first quadrant before moving on to all four quadrants of the coordinate plane.

Geometry

Taylor likes to draw different geometric figures on graph paper. If she plotted two vertices at (2, 3) and (6, 3), what two points might she use to complete a square? Use a graph to show how Taylor can prove that the figure she drew was a square.

Kaylee and Justin worked together to draw a parallelogram. How might they have completed the figure if they were given the two vertices (5, 8) and (10, 8)? Use a graph to show how they might prove their figure is a parallelogram.

Victor plans to draw two trapezoids on his graph paper. He intends to draw a large one with vertices at (–6, 4), (6, 4), (–4, –4), and (4, –4) and a smaller one inside the larger one. Where might Victor locate the vertices for the smaller trapezoid if the area is one-fourth the larger trapezoid.

Geometry

Flips, Slides, and Turns

Standard
Uses motion geometry to understand geometric relationships

Overview
These problems focus on geometric transformations.

Problem-Solving Strategies
- Act it out or use manipulatives
- Organize information in a picture, list, table, graph, or diagram

Materials
- *Flips, Slides, and Turns* (page 123; flipsslidesturns.pdf)
- *Motion Geometry* (motiongeo.pdf)
- *Student Response Form* (page 132; studentresponse.pdf) *(optional)*

Activate
1. Have students explain to you what it means for something to rotate.

2. Distribute copies of *Motion Geometry* to students. Discuss with the class that when a rotation occurs, it is always around a point. Ask students to rotate the first arrow 90°. Ask them to sketch the result. Ask them to rotate the same arrow 180° and to sketch their results.

3. Ask students to rotate the second arrow 90°. Ask them to sketch the result. Ask them to rotate the same arrow 180° and to sketch their results.

4. Ask students to rotate the third arrow 90°. Ask them to sketch the result. Ask them to rotate the same arrow 180° and to sketch their results.

5. Ask students to compare their sketches. What can they say about the results? What difference does the location of the point around which a rotation occurs make?

Solve
1. Distribute copies of *Flips, Slides, and Turns* to students. Have students work alone, in pairs, or in small groups.

2. As you observe students at work, watch for students who may rotate the figures clockwise instead of counterclockwise, which is the convention. Discuss the difference the direction of the rotation makes.

3. Invite several students to share their responses.

Debrief
1. How do you know the way in which to move the figure?

2. Does it matter if you reflect a figure before you rotate it or rotate it before reflecting it?

Differentiate

It is helpful for students who do not "see" what is happening to punch a small hole at the point of rotation and actually rotate the figure on the point of a pencil or compass to model the rotations.

Geometry

How might Lola reflect the following figure across the x-axis? Draw the result.

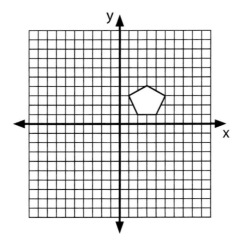

How might Linda rotate the following figure 90°? Draw the result.

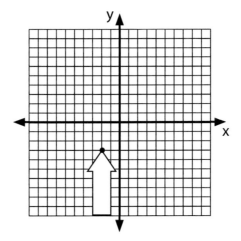

How might Bruce reflect the following figure across the x-axis, then rotate the figure 90°? Draw the result.

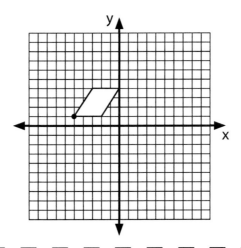

Geometry

What Is the Angle?

Standard
Understands characteristics of lines and angles

Overview
Students identify different types of angles, estimate the measure of angles, and compare angles.

Problem-Solving Strategies
- Guess and check or make an estimate
- Use logical reasoning

Materials
- *What Is the Angle?* (page 125; whatangle.pdf)
- *Angles* (angles.pdf)
- *Student Response Form* (page 132; studentresponse.pdf) *(optional)*

Activate
1. Have students show you what a 90° angle looks like using their thumbs and forefingers. Have students show you what a 180° angle looks like with their arms.
2. Distribute copies of *Angles* and direct students to draw a 90° angle at each vertex.
3. Challenge students to use the 90° angle to estimate the angle measure of the given angle.

Solve
1. Distribute copies of *What Is the Angle?* to students. Have students work alone, in pairs, or in small groups.
2. As students work, observe the strategy students are using. How effectively are students using the benchmark angle measures of 90° and 180°?

Debrief
1. How did you decide whether to use the benchmark angles of 90° or 180°?
2. Does the direction of the ray rotated on its axis make a difference when identifying the angle measure? If so, how?

Differentiate
Many students believe that the length of the rays that comprise the angle influence the actual angle measure. These students need more experiences working with angles with various ray measurements to dispel this misconception. Also, some students need to be encouraged to draw right angles on given figures to assist them in determining how close to 90° a given angle might be.

Geometry

Which of the following angles might Tracey label as right, acute, obtuse, or straight?

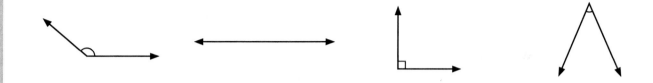

Devin is going to estimate the measures of the following angles, then order them from least to greatest. What are some estimates he might make? Justify your response.

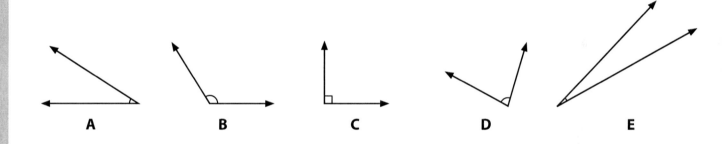

A   B   C   D   E

Mark drew the following map. Help him classify each angle created by the intersections. Estimate each angle measure.

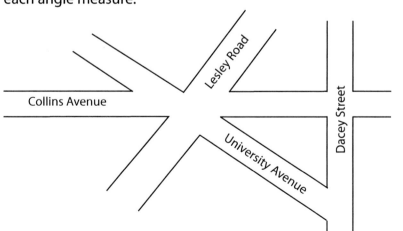

1. Collins Avenue and Dacey Street

2. University Avenue and Dacey Street

3. Lesley Road and University Avenue

4. University Avenue and Collins Avenue

5. Lesley Road and Collins Avenue

© Shell Education   #50777—50 Leveled Math Problems, Level 5   125

Geometry

Sort It Out

Standards
- Knows basic geometric language for describing and naming shapes
- Understands basic properties of figures

Overview
These problems focus on properties of geometric figures.

Problem-Solving Strategies
- Act it out or use manipulatives
- Organize information in a picture, list, table, graph, or diagram

Materials
- *Sort It Out* (page 127; sortout.pdf)
- *Student Response Form* (page 132; studentresponse.pdf) *(optional)*

Activate
1. Have students name as many polygons as they can.
2. Ask students to identify the characteristics of polygons. Write the list on the board.
3. Have students draw examples of polygons and non-polygons based on the list of characteristics.

Solve
1. Distribute copies of *Sort It Out* to students. Have students work alone, in pairs, or in small groups. Direct students to think about how they can prove their responses.
2. Have students share their sketches and matches for the overlapping figures and name both figures.

Debrief
1. How did you decide which two figures overlapped?
2. With which of the definitions did you have the most difficulty?
3. How did you decide whether all rectangles are squares or all squares are rectangles?

Differentiate ○ ★
For students who struggle to match the definition to the correct figure name, suggest they make a table listing specific characteristics, such as: 4 sides, 4 right angles, 1 right angle, 2 pairs of parallel sides, 1 pair of parallel sides, 4 congruent sides, 2 congruent sides. Students then fill in the name of each figure in the correct columns. So, for instance, the term *square* would be listed in the columns entitled *4 sides, 4 right angles, 2 pairs of parallel sides, 4 congruent sides*.

Geometry

Gregory drew the following outlines. Each was made by overlapping two figures. Sketch in both figures and name them.

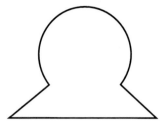   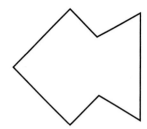   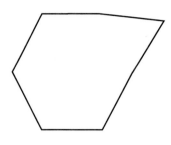

_____   _____   _____

_____   _____   _____

Rufus spilled his juice on a geometry story leaving a lot of blank spaces. Help him rewrite the story using the following terms: polygon, quadrilateral, parallelogram, trapezoid, rectangle, square, rhombus.

I am a special _____ because all my sides are congruent, not just two. I am called a _____ even though I think I am a pretty cool shape. I am also called a _____ because all my sides are straight lines and are joined at a vertex. I have a special cousin which some people call a diamond but which is really a _____. My cousins and I are all _____(s) because we all have four sides. Some people confuse two of my cousins. My favorite cousin often has two slanted sides and is called a _____, but it could also have two straight sides. My most confusing cousin is a _____ and has at least one pair of parallel sides.

Ali completed the following phrases. Her teacher told her she made many mistakes. Help Ali correct her work. Write in the correct phrase.

can be	is always	is never

1. A parallelogram is never a square. _____

2. A rectangle is always a square. _____

3. A trapezoid is always a square. _____

4. A rhombus is never a square. _____

5. A quadrilateral is always a square. _____

Geometry

Geometric Nets

Standard
Understands basic properties of figures

Overview
Students use two-dimensional nets to represent three-dimensional figures.

Problem-Solving Strategies
- Act it out or use manipulatives
- Organize information in a picture, list, table, graph, or diagram

Materials
- *Geometric Nets* (page 129; geonets.pdf)
- boxes of various sizes including a cube, rectangular prism, cylinder, and a triangular prism
- scissors
- graph paper (graphpaper.pdf)
- *Student Response Form* (page 132; studentresponse.pdf) *(optional)*

Activate
1. Have students explain to you everything they know about the relationship between a two-dimensional net and a three-dimensional object. Record their comments on the conjecture board.
2. Hold up the box that is a cube. Challenge students to sketch what the cube would look like if it were cut apart and shown as a two-dimensional figure.
3. Ask several volunteers to share their sketches. Then, cut the box edges to show the two-dimensional net.

Solve
1. Distribute copies of *Geometric Nets* to students. Have students work alone, in pairs, or in small groups.
2. Direct students to sketch either the two-dimensional nets or the three-dimensional objects as directed.

Debrief
1. How can you decide if the two-dimensional net matches the three-dimensional object?
2. Does it matter where you place the height of the object?
3. Is there more than one way to draw a net for a three-dimensional object? How do you know?

Differentiate ○ ■ △ ☆
Kinesthetic learners may benefit from making nets to model the nets. Allow students to model the nets on graph paper, cut them out, then fold them to determine if the net actually makes the desired object.

Geometry

How might Angie draw the three-dimensional object represented by the two-dimensional net shown below? What is the shape called?

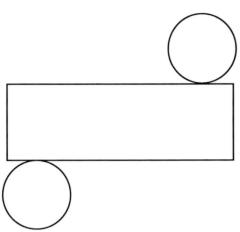

How might Emmett draw the two-dimensional net for the three-dimensional object shown below?

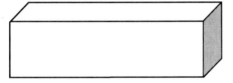

Reese claims it is possible to draw ten different two-dimensional nets for a cube. Can you find all ten?

Geometry

Graph It

Standard
Knows basic characteristics and features of the rectangular coordinate system

Overview
Students graph points on the coordinate plane.

Problem-Solving Strategies
- Act it out or use manipulatives
- Organize information in a picture, list, table, graph, or diagram

Materials
- *Graph It* (page 131; graphit.pdf)
- *Paper and Pencil Battleship* (battleship.pdf)
- graph paper (graphpaper.pdf)
- *Student Response Form* (page 132; studentresponse.pdf) *(optional)*

Activate
1. Distribute *Paper and Pencil Battleship* and graph paper to students. Have students draw and label a coordinate plane up to 10 or –10 on all axes.
2. Explain the rules of the game, and model the game for the class.
3. Allow students to play the game with their partners.

Solve
1. Distribute copies of *Graph It* to students. Have students work alone, in pairs, or in small groups.
2. As students work, observe whether they are plotting the ordered pairs at the correct locations and are connecting the points in each row.
3. Invite a student to share the joke together with the answer, or the saying. Ask if anyone got a different answer or did it differently.

Debrief
1. What did you find was the most challenging?
2. What suggestions might you give another student about graphing?

Differentiate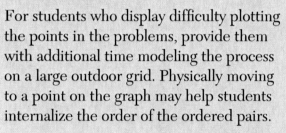
For students who display difficulty plotting the points in the problems, provide them with additional time modeling the process on a large outdoor grid. Physically moving to a point on the graph may help students internalize the order of the ordered pairs.

Geometry

Nolan is playing a game with his partner. He hid a message in his graph. Graph the ordered pairs to find his message. What is the message he sent to his friend?

- (2, 13), (0, 13), (0, 17), (2, 17)
- (9, 15), (10, 15)
- (9, 5), (10, 3)
- (2, 3), (2, 7), (4, 7), (4, 5), (2, 5)
- (6, 13), (6, 17), (7, 14), (8, 17), (8, 13)
- (11, 7), (13, 7)
- (17, 13), (17, 17), (19, 17), (19, 13), (17, 13)
- (5, 3), (5, 5), (6, 7), (7, 5), (7, 3)
- (15, 5), (15, 3)
- (3, 13), (3, 17), (5, 17), (5, 13), (3, 13)

- (5, 5), (7, 5)
- (14, 7), (15, 5), (16, 7)
- (11, 13), (9, 13), (9, 17), (11, 17)
- (14, 17), (16, 17)
- (7, 8), (7, 12), (8, 9), (9, 12), (9, 8)
- (12, 7), (12, 3)
- (8, 3), (8, 7), (10, 7), (10, 5), (8, 5)
- (10, 12), (11, 10), (12, 12)
- (11, 10), (11, 8)
- (15, 17), (15, 13)

Why did the scientist visit the mathematician's house? Graph the ordered pairs to reveal the answer. Lift your pencil after each row.

- (5, 8), (3, 8), (3, 12), (5, 12)
- (10, 7), (12, 7)
- (2, 13), (2, 17)
- (12, 8), (14, 8)
- (15, 8), (15, 12), (17, 8), (17, 12)
- (9, 12), (10, 8), (11, 12)
- (13, 13), (13, 15), (14, 17), (15, 15), (15, 13)
- (7, 10), (8, 8)

- (2, 15), (4, 15)
- (20, 12), (18, 11), (18, 9), (20, 8), (20, 10), (19, 10)
- (13, 15), (15, 15)
- (6, 8), (6, 12), (8, 12), (8, 10), (6, 10)
- (16, 13), (18, 13), (18, 15), (16, 15), (16, 17), (18, 17)
- (4, 13), (4, 17)
- (5, 15), (6, 15)
- (12, 12), (14, 12)

- (7, 3), (7, 7), (9, 7), (9, 5), (7, 5)
- (7, 13), (5, 13), (5, 17), (7, 17)
- (3, 10), (4, 10)
- (13, 8), (13, 12)
- (10, 3), (12, 3)
- (10, 17), (10, 13), (11, 15), (12, 13), (12, 17)
- (11, 3), (11, 7)
- (0, 8), (2, 8), (2, 10), (0, 10), (0, 12), (2, 12)

Carter hid a message on the coordinate plane. He is challenging you to plot the points to find his message. Connect each point to find his message.

- (−6, 1), (−6, 5), (−5, 2), (−4, 5), (−4, 1)
- (5, 1), (5, 5)
- (1, 5), (1, 1)
- (−3, 3), (−1, 3)
- (1, −6), (−1, −6), (−1, −2), (1, −2)
- (4, −2), (2, −4), (4, −6)
- (3, 3), (5, 3)
- (2, −6), (2, −2)

- (−4, −6), (−4, −2), (−2, −2), (−2, −6), (−4, −6)
- (−3, 1), (−3, 3), (−2, 5), (−1, 3), (−1, 1)
- (5, −6), (7, −6), (7, −4), (5, −4), (5, −2), (7, −2)
- (−6, −4), (−5, −6)
- (3, 1), (3, 5)
- (0, 5), (2, 5)
- (−7, −6), (−7, −2), (−5, −2), (−5, −4), (−7, −4)

Appendix A

Name: _____ Date: _____

Student Response Form

Problem:

(glue your problem here)

My Work and Illustrations:
(picture, table, list, graph)

My Solution:

My Explanation:
_____
_____
_____

Appendix B

Name: _____ Date: _____

Observation Form

	Criteria	Notes
Communication	Variety of Methods (written, oral, etc.)	
	Interaction with Peers • Are they on topic? • Are they respectful?	
	Teacher/Student How does student respond to teacher inquiry?	
	Clarifying Questions What types of questions are being asked within the group?	
	Makes Connections • Problem to other math concepts • Problem to real world • Problem to other content	
Problem Solving	Chooses Appropriate Strategies • What strategy did they choose? • Is the strategy efficient?	
	Reasonableness Does the answer make sense numerically and contextually?	
	Confidence in approaching problem	
Reasoning and Proof	Defend and Justify Can they mathematically defend and justify their thinking?	
	Inductive vs. Deductive • Inductive—looks for patterns and makes generalizations • Deductive—makes logical arguments, draws conclusions, applies generalizations to specific situations	
Content	Accuracy • How accurate is the work being done? • Are there any particular mistakes being made?	

© Shell Education

Record-Keeping Chart

Use this chart to record the problems that were completed. Record the name of the lesson and the date when the appropriate level was completed.

Name: _____

Lesson	⬤ Date Completed	◼ Date Completed	▲ Date Completed

Answer Key

In What Order? (page 33)

● a. She is not correct.; 7 − 5 + 6; 2 + 6; 8
b. She is not correct.; 8 + 2 × 2; 8 + 4; 12

■ a. 3 + (3 × 3) − 5 × 2; 3 + 9 − 5 × 2; 3 + 9 − 10; 12 − 10; 2
b. 10 ÷ 5 × (2 + 60) − 15; 10 ÷ 5 × 62 − 15; 2 × 62 − 15; 124 − 15; 109

▲ a. (4 + 4) × 4 + 4 = 36
b. (4 + 4 + 4) × 4 = 48 or (4 × 4 − 4) × 4 = 48

Order Counts (page 35)

● a. 18 − 6 + 10   b. 28 − 10 + 3
■ a. 9 + 24 − 4   b. 9 ÷ 3 × 2
▲ a. 22 − 2   b. 104 + 4

Number Patterns (page 37)

● 35, 42, 49, 56; 70, 84, 98, 112; The second pattern is double the first pattern.

■ 19, 21, 23, 25; 57, 63, 69, 75; The second pattern is three times the first pattern.

▲ 20, 40, 50, 70, 80; 6, 8, 10, 14, 18; The second pattern is one-fifth of the first pattern; New patterns will vary.

Geometric Patterns (page 39)

● 1.
2. 12 squares
3. x + 2

■ 1.
2. 42 Xs
3. 2x + 2

▲ 1.
2. 10,000 dots
3. x^2

How Else Might I Look? (page 41)

● 4 × 8 + 3 × 8 = (4 + 3) × 8
■ 90 + 75; 12 × 90; 26 × 11
▲ The first expressions are not equivalent because multiplying 3 by 78 then adding 4 is not the same as multiplying 4 by 78 then adding 3. The second expressions are equivalent due to the distributive property.

Where Am I? (page 43)

●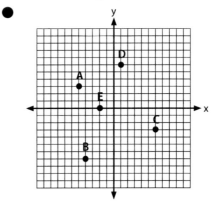

■ 18 blocks
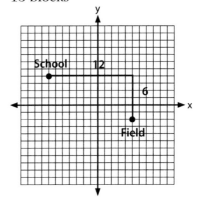

▲ 43 blocks
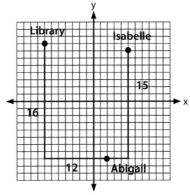

Answer Key (cont.)

How Do I Change? (page 45)

- $y = 2x + 1$

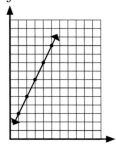

- 9, 21; $y = 4x + 1$

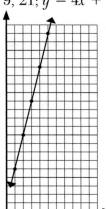

- 3, 16, 5; $y = x^2$

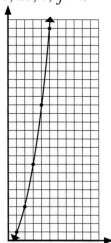

What's Our Relation? (page 47)

- 10, 22, 38, 50, 66; $y = 4x + 6$
- 49, 98, 7n; $y = 7n$; 98, 196, 294, 14n; $y = 14n$
- Jessa will raise more money.; $y = \$1.50x$; $y = \$0.75x + \$3.00$

Name My Number (page 49)

- 2,016,435
- 718,925,634
- 483,917,526

Rectangular Products (page 51)

- 920 stamps
- $216
- $120 \times 40 + 120 \times 5$ or $100 \times 45 + 20 \times 45$; 5,400 sq. ft.

Whatever Remains (page 53)

- 5 vans
- $6.25; No.
- No, he needs 10 more.

Grouping or Sharing? (page 55)

- 268 soldiers
- Answers will vary.
- $9,765 \div 46$

Dealing with Decimals (page 57)

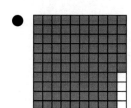

- $0.90
- $3.22

Expanded Form (page 59)

- $6{,}000 + 800 + 10 + 5$ or $6 \times 10^3 + 8 \times 10^2 + 1 \times 10^1 + 5 \times 10^0$; $7 + 0.6 + 0.09 + 0.008$ or $7 \times 10^0 + 6 \times 10^{-1} + 9 \times 10^{-2} + 8 \times 10^{-3}$
- 20,506; 5,109
- $7 \times 10^4 + 5 \times 10^3 + 3 \times 10^2 + 9 \times 10^1 + 7 \times 10^0 + 6 \times 10^{-2} + 3 \times 10^{-3}$; 40,730.52

Answer Key (cont.)

Travel Expenses (page 61)

- ● $14.44
- ■ $1,172.25; $27.75
- ▲ 6 vans; $172.00

Computing with Decimals (page 63)

- ● 0.63 sq. cm; $A = 0.7 \times 0.9$
- ■ 20.125 sq. in.
- ▲ ○ = 0.5, □ = 0.8, △ = 0.7

Dizzying Decimals (page 65)

- ● $0.60
- ■ 249 bottles
- ▲ 28 grandchildren

Estimating Decimals (page 67)

- ● $200 \div 80 = 2.5$
- ■ $200 \div 10 = 20$ lbs.; about 18 lbs.; about 2 lbs. over
- ▲ $(\$200 + \$40) \div 20 = \$12$; $13.51; $1.51 under

About How Much? (page 69)

- ● No; 150
- ■ Use front-end estimation. $4 + 80 + 10 + 400 = 494$
- ▲ Kelly's estimate is closer. $40 + 200 + 70 + 100 + 10 = 420$

Where Do I Fit? (page 71)

- ● $\frac{5}{7}$;
- ■ Olivia has more beads left.; Explanations will vary.
- ▲ $\frac{1}{8}, \frac{3}{8}, \frac{5}{8}, \frac{2}{3}, \frac{4}{5}, \frac{5}{6}$; Explanations will vary.

Ribbons and Bows (page 73)

- ● $\frac{10}{12}$ ft.
- ■ $5\frac{3}{4}$ in., $5\frac{1}{2}$ in., $5\frac{3}{8}$ in., $5\frac{1}{4}$ in., $5\frac{1}{8}$ in.
- ▲ $3\frac{1}{4}$ yds, $3\frac{5}{16}$ yds, $3\frac{1}{2}$ yds, $3\frac{9}{16}$ yds, $3\frac{5}{8}$ yds, $3\frac{3}{4}$ yds, $3\frac{13}{16}$ yds, $3\frac{7}{8}$ yds

Fractional Sums (page 75)

- ● $\frac{7}{8}$ c.
- ■ $2\frac{5}{12}$ hours
- ▲ Andrew is correct.

What's the Difference? (page 77)

- ● a. $\frac{9}{28}$
  b. $1\frac{1}{15}$
  c. $4\frac{5}{8}$
  d. $1\frac{1}{10}$
- ■ $3\frac{5}{8}$ pizzas
- ▲ $9\frac{5}{24}$ hours

It's Close To What? (page 79)

- ● Explanations will vary.
- ■ No, he is incorrect. $\frac{2}{4}$ is a half and $\frac{2}{3}$ is greater than $\frac{1}{2}$ so the sum is greater than 1 whole.
- ▲ $\frac{1}{3} + \frac{1}{4}$ is close to $\frac{1}{2}$ and $\frac{5}{6}$ is greater than $\frac{1}{2}$ so her estimate is greater than 1 whole pie.

More or Less (page 81)

- ● $\frac{5}{12}$ hours
- ■ $1\frac{5}{8}$ miles
- ▲ $19\frac{11}{12}$ hours; $40\frac{1}{12}$ hours

Fractional Areas (page 83)

- ● A: $\frac{1}{2}$; B: $\frac{1}{8}$; C: $\frac{1}{8}$; D: $\frac{1}{8}$; E: $\frac{1}{8}$
- ■ A: $\frac{1}{4}$; B: $\frac{1}{16}$; C: $\frac{1}{16}$; D: $\frac{1}{16}$; E: $\frac{1}{16}$; F: $\frac{1}{4}$; G: $\frac{1}{16}$; H: $\frac{1}{16}$; I: $\frac{1}{16}$; J: $\frac{1}{16}$
- ▲ A: $\frac{1}{4}$; B: $\frac{1}{8}$; C: $\frac{1}{8}$; D: $\frac{3}{8}$; E: $\frac{1}{16}$; F: $\frac{1}{32}$; G: $\frac{1}{32}$

Answer Key (cont.)

The Product Is Smaller (page 85)

- ● $\frac{2}{3} \times \frac{4}{5} = \frac{8}{15}$

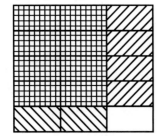

- ■ She took $\frac{1}{3}$ of the pan of brownies.

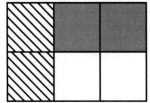

- ▲ $\frac{5}{6}$ lb.; $1\frac{2}{3}$ lbs.; Jill ate $\frac{5}{12}$ lbs. more strawberries.

Fair Sharing, Equal Groups (page 87)

- ● 7 bags
- ■ 16 students
- ▲ $3\frac{3}{4} \div \frac{1}{2} = 7\frac{1}{2}$ rolls

Map Reading (page 89)

- ● 72 miles
- ■ $33\frac{1}{3}$ miles
- ▲ 6 miles

Fill It Up (page 91)

- ● 336 blocks
- ■ 1 × 1 × 72; 1 × 3 × 24; 1 × 6 × 12; 1 × 8 × 9; 1 × 4 × 18; 1 × 2 × 36; 2 × 4 × 9; 2 × 2 × 18; 2 × 3 × 12; 2 × 6 × 6; 3 × 4 × 6; 3 × 3 × 8
- ▲ 1 × 1 × 36; 1 × 2 × 18; 1 × 3 × 12; 1 × 4 × 9; 1 × 6 × 6; 2 × 2 × 9; 2 × 3 × 6; 3 × 3 × 4; 3 × 3 × 4 is the most efficient box.

How Spacious Is It? (page 93)

- ● 360 square feet
- ■ A = 5 sq. units; A = 13 sq. units; A = 10 sq. units
- ▲ Answers will vary, but can include 602 ft., 304 ft., 206 ft., 130 ft., 112 ft., 80 ft., 74 ft., 70 ft.

Cubic Views (page 95)

- ● 6 layers
- ■ 240 cubes
- ▲ 7 in.

Volume in Practice (page 97)

- ● 1,440 cubic feet
- ■ 48 boxes
- ▲ 304 cubic feet

What's My Unit? (page 99)

- ● pint, quart, gallon
- ■ 8 km, 8 m, 8 dm, 8 cm, 8 mm
- ▲ ounce, pint, quart, gallon

Metrically Speaking (page 101)

- ● 1,000 mm
- ■ 500 dimes
- ▲ 9,500 m

How Much Is There? (page 103)

- ● Emma; 1,314 milliliters; Answers will vary.
- ■ 8.375 liters
- ▲ 625 milliliters; yes; 125 milliliters

Answer Key (cont.)

All in a Line (page 105)

- ```
  x           x           x
  x           x           x       x
  x   x x x x x       x       x
  14 15 16 17 18 19 20 21 22 23 24 25 26
  ```
- ■ median: 19 miles; mode: 14 miles, 18 miles, 23 miles; mean: 19.6 miles
- ▲ 19.6 miles

What Is the Favorite? (page 107)

- ● baseball: 50 students; swimming: 25 students; bowling: 25 students; football: 25 students; soccer: 50 students; lacrosse: 25 students
- ■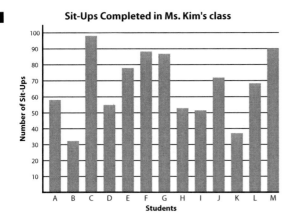
  mean: 66.3 sit-ups; Answers will vary.
- ▲
  Answers will vary.

Stem-and-Leaf Plots (page 109)

- ● ```
  3 | 5
  4 |
  5 | 7
  6 | 1 5 7 8 8
  7 | 3 3 4 5
  8 | 2
  ```
- ■ Answers will vary.
- ▲ mode: Eagles: 48, Pumas: 53; median: Eagles: 49.5, Pumas: 49; Answers will vary.

## What Does It Mean? (page 111)

- ● 83.17 points
- ■ average score
- ▲ 93

## The Plot Thickens (page 113)

- ● $\frac{1}{2}$ milliliters
- ■ $\frac{3}{8}$ milliliters
- ▲ $\frac{35}{72}$ milliliters

## One or the Other (page 115)

- ● $\frac{1}{2}$; $\frac{1}{2}$
- ■ $\frac{1}{2}$; $\frac{1}{2}$; 0; 1
- ▲ Answers will vary.

## Congruency (page 117)

- ●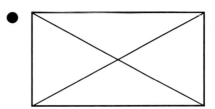
- ■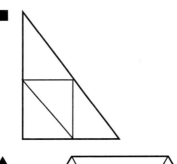
- ▲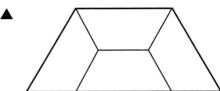

# Answer Key (cont.)

## Classifying Figures (page 119)

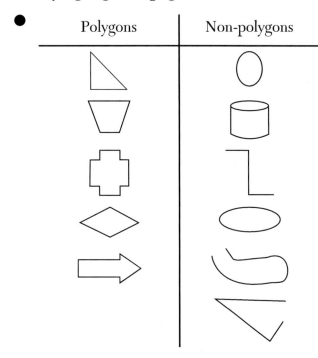

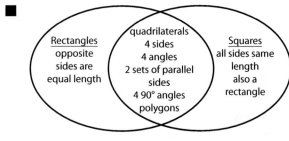

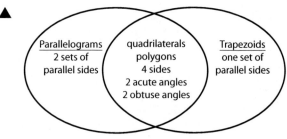

## Plots A Lot (page 121)

- ● (2, –1) and (6, –1) or (2, 7) and (6, 7)
- ■ (3, 4) and (8, 4) or (7, 4) and (12, 4)
- ▲ (3, 2), (–3, 2), (–1, –2), (1, –2)

## Flips, Sides, and Turns (page 123)

●
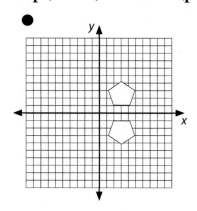

■
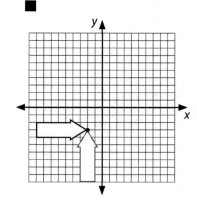

▲
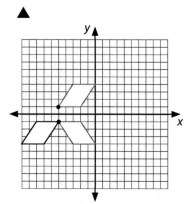

# Answer Key (cont.)

## What Is the Angle? (page 125)

- obtuse; straight; right; acute
- ■ E, A, D, C, B; Estimates will vary.
- ▲ 1. right; 90°
  2. acute; Estimates will vary.
  3. acute; Estimates will vary.
  4. obtuse; Estimates will vary.
  5. obtuse; Estimates will vary.

## Sort It Out (page 127)

- ●  circle and    square and    hexagon and
     trapezoid    trapezoid     triangle

 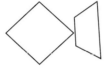 

- ■ rectangle; square; polygon; rhombus; quadrilateral; parallelogram; trapezoid
- ▲ 1. A parallelogram **can be** a square.
  2. A rectangle **can be** a square.
  3. A trapezoid **is never** a square.
  4. A rhombus **can be** a square.
  5. A quadrilateral **can be** a square.

## Geometric Nets (page 129)

- ● cylinder

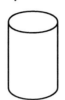

- ■

- ▲ There are 10 additional nets.

## Graph It (page 131)

- ● Come to my party
- ■ He was serving pi
- ▲ Math rocks

# References Cited

Bright, G. W., and J. M. Joyner. 2005. *Dynamic classroom assessment: Linking mathematical understanding to instruction.* Chicago, IL: ETA Cuisenaire.

Brown, S. I., and M. I. Walter. 2005. *The art of problem posing.* Mahwah, NJ: Lawrence Earlbaum.

Cai, J. 2010. Helping elementary students become successful mathematical problem solvers. In *Teaching and learning mathematics: Translating research for elementary school teachers*, ed. D. V. Lambdin and F. K. Lester, Jr., 9–13. Reston, VA: NCTM.

D'Ambrosio, B. 2003. Teaching mathematics through problem solving: A historical perspective. In *Teaching mathematics through problem solving: Prekindergarten–Grade 6*, ed. F. K. Lester, Jr. and R. I. Charles, 37–50. Reston, VA: NCTM.

Goldenberg, E. P., N. Shteingold, and N. Feurzeig. 2003. Mathematical habits of mind for young children. In *Teaching mathematics through problem solving: Prekindergarten–Grade 6*, ed. F. K. Lester, Jr. and R. I. Charles, 51–61. Reston, VA: NCTM.

Michaels, S., C. O'Connor, and L. B. Resnick. 2008. Deliberative discourse idealized and realized: Accountable talk in the classroom and in civil life. *Studies in philosophy and education* 27 (4): 283–297.

National Center for Educational Statistics. 2010. Highlights from PISA 2009: Performance of U.S. 15-year-old students in reading, mathematics, and science literacy in an international context. http://nces.ed.gov/pubsearch/pubsinfo.asp?pubid=2011004

National Governors Association Center for Best Practices and Council of Chief State School Officers. 2010. Common core state standards. http://www.corestandards.org/the-standards.

National Mathematics Advisory Panel. 2008. *Foundations for success: The final report of the National Mathematics Advisory Panel.* Washington, DC: U.S. Department of Education.

Polya, G. 1945. *How to solve it: A new aspect of mathematical method.* Princeton, NJ: Princeton University Press.

Sylwester, R. 2003. *A biological brain in a cultural classroom.* Thousand Oaks, CA: Corwin Press.

Tomlinson, C. A. 2003. *Fulfilling the promise of the differentiated classroom: Strategies and tools for responsive teaching.* Alexandria, VA: ASCD.

Vygotsky, L. 1986. *Thought and language.* Cambridge, MA: MIT Press.

Appendix F

# Contents of the Teacher Resource CD

## Teacher Resources

| Page | Resource | Filename |
|---|---|---|
| 27–31 | Common Core State Standards Correlation | ccss.pdf |
| N/A | NCTM Standards Correlation | nctm.pdf |
| N/A | TESOL Standards Correlation | tesol.pdf |
| N/A | McREL Standards Correlation | mcrel.pdf |
| 132 | Student Response Form | studentresponse.pdf |
| 133 | Observation Form | obs.pdf |
| 134 | Record-Keeping Chart | recordkeeping.pdf |

## Lesson Resource Pages

| Page | Lesson | Filename |
|---|---|---|
| 33 | In What Order? | whatorder.pdf |
| 35 | Order Counts | ordercounts.pdf |
| 37 | Number Patterns | numberpatterns.pdf |
| 39 | Geometric Patterns | geopatterns.pdf |
| 41 | How Else Might I Look? | mightlook.pdf |
| 43 | Where Am I? | whereami.pdf |
| 45 | How Do I Change? | change.pdf |
| 47 | What's Our Relation? | relation.pdf |
| 49 | Name My Number | namenumber.pdf |
| 51 | Rectangular Products | rectproducts.pdf |
| 53 | Whatever Remains | remains.pdf |
| 55 | Grouping or Sharing? | grouping.pdf |
| 57 | Dealing with Decimals | dealdecimals.pdf |
| 59 | Expanded Form | expandedform.pdf |
| 61 | Travel Expenses | expenses.pdf |
| 63 | Computing with Decimals | computing.pdf |
| 65 | Dizzying Decimals | dizzying.pdf |
| 67 | Estimating Decimals | estimatingdec.pdf |
| 69 | About How Much? | abouthowmuch.pdf |
| 71 | Where Do I Fit? | wherefit.pdf |
| 73 | Ribbons and Bows | ribbonsbows.pdf |
| 75 | Fractional Sums | fractionalsums.pdf |
| 77 | What's the Difference? | difference.pdf |
| 79 | It's Close to What? | closetowhat.pdf |
| 81 | More or Less | moreless.pdf |
| 83 | Fractional Areas | fractionalareas.pdf |

| Page | Lesson | Filename |
|---|---|---|
| 85 | The Product Is Smaller | productsmaller.pdf |
| 87 | Fair Sharing, Equal Groups | fairequal.pdf |
| 89 | Map Reading | mapreading.pdf |
| 91 | Fill It Up | fillup.pdf |
| 93 | How Spacious Is It? | spacious.pdf |
| 95 | Cubic Views | cubicviews.pdf |
| 97 | Volume in Practice | volpractice.pdf |
| 99 | What's My Unit | whatunit.pdf |
| 101 | Metrically Speaking | metrically.pdf |
| 103 | How Much Is There? | howmuchthere.pdf |
| 105 | All in a Line | allinaline.pdf |
| 107 | What Is the Favorite? | favorite.pdf |
| 109 | Stem-and-Leaf Plots | stemleaf.pdf |
| 111 | What Does It Mean? | whatmean.pdf |
| 113 | The Plot Thickens | plotthickens.pdf |
| 115 | One or the Other | oneother.pdf |
| 117 | Congruency | congruency.pdf |
| 119 | Classifying Figures | classifying.pdf |
| 121 | Plots A Lot | plotsalot.pdf |
| 123 | Flips, Slides, and Turns | flipsslidesturns.pdf |
| 125 | What Is the Angle? | whatangle.pdf |
| 127 | Sort It Out | sortout.pdf |
| 129 | Geometric Nets | geonets.pdf |
| 131 | Graph It | graphit.pdf |

Appendix F

# Contents of the Teacher Resource CD (cont.)

## Additional Lesson Resources

| Page | Lesson | Filename |
|---|---|---|
| 32 | Order of Operations Hopscotch | hopscotch.pdf |
| 34 | Practice Expressions | expressions.pdf |
| 36 | Pattern Examples | patternexamples.pdf |
| 40 | Expression Practice | expressionprac.pdf |
| 36, 42, 44, 50, 54, 90, 92, 120, 128, 130 | graph paper | graphpaper.pdf |
| 48 | Place Value Number | placevalue.pdf |
| 50 | Array Model | arraymodel.pdf |
| 56, 62, 64 | Decimal Grids | decimalgrids.pdf |
| 74 | Fraction Strips | fractionstrips.pdf |
| 80 | Number Line | numberline.pdf |
| 88 | Maps | maps.pdf |
| 90, 94 | dot paper | dotpaper.pdf |
| 92 | Dot Paper Polygons | dotpolygons.pdf |
| 100 | Metric Conversion | conversion.pdf |
| 116 | Congruent Figures | congruentfigs.pdf |
| 118 | Classifying Geometric Figures | geofigures.pdf |
| 122 | Motion Geometry | motiongeo.pdf |
| 124 | Angles | angles.pdf |
| 130 | Paper and Pencil Battleship | battleship.pdf |